Lecture Notes
in Control and Information Sciences 317

Editors: M. Thoma · M. Morari

Chuan Ma · W. Murray Wonham

Nonblocking Supervisory Control of State Tree Structures

With 114 Figures

Springer

Authors

Dr. Chuan Ma
Prof. W. Murray Wonham
University of Toronto
Dept. of Electrical and Computer Engineering
10 King's College Road
Toronto, Ontario
Canada M5S 3G4

ISSN 0170-8643

ISBN-10 3-540-25069-7 **Springer Berlin Heidelberg New York**
ISBN-13 978-3-540-25069-2 **Springer Berlin Heidelberg New York**

Library of Congress Control Number: 2005923293

Springer is a part of Springer Science+Business Media

springeronline.com

Typesetting: Data conversion by author.
Final processing by PTP-Berlin Protago-TEX-Production GmbH, Germany
Cover-Design: design & production GmbH, Heidelberg
Printed on acid-free paper 89/3141/Yu - 5 4 3 2 1 0

Preface

Supervisory control refers to control of discrete-event systems. In such a system the logical order of the events is of concern and not the time at which the events take place. Examples of engineering control problems for which supervisory control is useful include: control of heating and ventilation systems, failure diagnosis of communication networks, design of logical controllers in automated cars, control of a metro network, etc. The RW-framework developed by P.J. Ramadge and W.M. Wonham in terms of generators, directly related to automata, is now widely used. There exists now a body of theory on the existence of a supervisor which achieves control objectives of safety and required behavior, and for their computation. A bottleneck for the application of the theory to engineering examples is the large computational complexity of controllers. Approaches to lower the complexity are modular and hierarchical control. The existing algorithms for modular control in general still require a large complexity. Hence the interest in hierarchical control.

A formalism for hierarchical systems and for their model-checking was proposed by David Harel in a paper published in 1987. For control of discrete-event systems a similar approach was developed by W.M. Wonham and his Ph.D. students but the computational and complexity issues were still formidable. This research program is continued with this book. The plan of attack, constructed in hindsight, specified: (1) a new formalism for supervisory control of hierarchical systems; (2) specifications formulated in terms of logical formulas so as to lower their computational complexity; and (3) the use of algorithms based on binary decision diagrams (BDDs). This plan has been successfully carried out by Chuan Ma during his Ph.D. study under supervision of W. Murray Wonham. The results of this book constitute a major advance for supervisory control. Yet, more research and experience is required to make the theory and algorithms into an effective tool for control engineering of discrete-event systems.

The undersigned highly recommends the reading of this book to researchers in control of discrete-event systems. The authors are congratulated with the appearance of this book!

Amsterdam, 24 January 2005 *Jan H. van Schuppen*

Contents

List of Figures

List of Tables

1

Introduction

In control theory, the dynamic behavior of Continuous-State Systems (CSS) is typically modelled by differential or difference equations. A considerable number of analysis and synthesis techniques have been developed for CSS on this basis. In recent years, modern technologies, especially computer technology, have fuelled the creation of complex man-made Discrete-Event Systems (DES). A DES is discrete in time and in state space, and event-driven rather than time-driven. Typical examples can be found in flexible manufacturing systems, communication protocols, embedded reactive systems, traffic control systems, multi-agent systems and military C3I systems. Here a DES can no longer be formally described by differential or difference equations, and therefore new control theories must be investigated.

A number of DES control theories have been established. However, no theory can be singled out as dominating. A survey of some DES theories is given in [Ho89]. The present monograph, adapted from [Ma04], is an extension to the RW Supervisory Control Theory (SCT) of DES.

1.1 RW Supervisory Control Theory and Architectures

The RW SCT [1], initiated by Ramadge and Wonham (RW) [RW82, Ram83, RW87a], is the first control theory for a general class of DES. In the original RW framework, a DES (plant) is modelled as an *automaton* **G**, and its behavior is described as a *formal language* $L(\mathbf{G})$ generated by the automaton over an alphabet Σ of event labels. Σ is partitioned into two subsets: the set Σ_c of *controllable* events that can be disabled by an external controller and the set Σ_u of *uncontrollable* events that cannot. With this control mechanism, a controller can restrict the system behavior only by disabling controllable events. We call such a controller a *supervisor*. Supervision can be implemented by the feedback configuration in Figure 1.1, where the supervisor exercises con-

[1] For an extensive introduction to SCT, please review [Won04]. Available at `http://www.control.utoronto.ca/DES`.

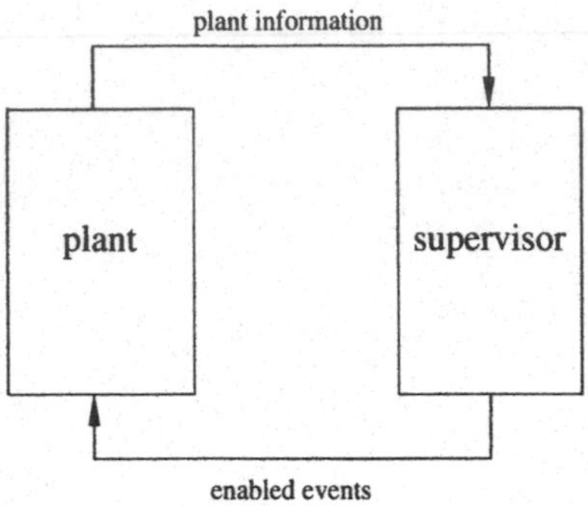

Fig. 1.1. Supervision Implementation

trol actions based on the past behavior of the controlled plant. The original RW framework is language-based. So the "plant information" in Figure 1.1 is given by a *string* generated by the plant automaton up to the current moment. After [RW87b, LW88, LW93] introduced the *predicate* concept into the RW framework, the "plant information" in Figure 1.1 can also be the current state of the plant automaton (possibly augmented by memory units).

Given the desired behavior as a formal language E, a supervisor can be designed to restrict the plant behavior such that the controlled system behavior is a sublanguage of E. It is often of interest to obtain an *optimal* supervisor, where "optimal" means "minimally restrictive". Also it is of interest to have a *nonblocking* supervisor, under which there is always the possibility, although not a guarantee, for the controlled system to eventually complete predefined critical tasks.

In the RW framework, the synthesis of an optimal nonblocking supervisor plays a central role. The first algorithm was given in [WR87]. Its computational complexity is of polynomial order in the model's state size. However, a complex DES is often modelled as the product of a number of simpler components. So its state size increases exponentially with the number of components. Later, Gohari and Wonham [GW00] proved that the synthesis problem is in fact NP-hard: *specifically it is unlikely that any algorithm can be found to solve it that circumvents state space explosion that is exponential in the number of system components.*

It is therefore attractive to explore structured system architectures with the property that, if the given system can be modelled in the selected framework, the required computation can be carried out with greater efficiency than by a naive 'monolithic' approach. So far, there are two kinds of well known architectures: *horizontal* and *vertical* modularity. By exploiting modularity, several control strategies have been proposed.

1. There are two methodologies with horizontal modularity. In [WR88], a modular approach was proposed. The basic idea is to divide the overall synthesis problem into several simpler problems. Conditions were also

given to guarantee the nonblocking property of the controlled system. [LW90], extended by [RW92, BL00], proposed decentralized control of DES, which is similar to modular control except that each supervisor can access only partial information and is allowed to disable only a subset of controllable events.

2. The methodology of exploiting vertical modularity is usually called hierarchical control. Zhong and Wonham [ZW90] studied a two-level hierarchy consisting of a low-level DES and a high-level aggregate DES. Only the high-level DES is involved in the synthesis for a given high-level specification. A fundamental concept in hierarchical control, "hierarchical consistency", was introduced to ensure that the high-level supervisor can actually be implemented in the low-level. [WW96] generalized Zhong's hierarchical control, based on the concepts of control structures and observers, and Zhong's hierarchical consistency can be achieved from a new condition called "control consistency". This kind of hierarchical control is bottom-up in the sense that the hierarchical consistency is built from the low-level model to the high-level. Hubbard and Caines [HC02] presented a state-based approach to this bottom-up hierarchy, where "trace-dynamical consistency" was introduced to ensure hierarchical consistency.

Apart from the effort needed to adapt a specific structural form to the system to be modelled, the price to be paid may include relaxing the (monolithic) requirement of optimality, viz. maximal permissiveness of controlled behavior. However, as long as controlled behavior is 'legal' and nonblocking, some reduction in permissiveness may well be acceptable for the sake of increased tractability. In addition, structured modelling may confer advantages of model transparency and modifiability. One instance is Leduc's [Led02, LBW01, LLW01b, LLW01c] *Hierarchical Interface-based Supervisory Control* theory (HISC). Another one is Shen's [SC02] *Hierarchically Accelerated Dynamic Programming* (HADP).

1.2 Motivation and Related Works

In the course of synthesis, our fundamental procedures are various set operations performed on the model's state space. Among them are intersection, union, and complement. If we store these state sets in the computer by enumerating all of their elements (as handled in TCT [2]), we face the state explosion problem head on. That is, the computational complexity of the required set operations is of polynomial order in the model's state size.

So in the first step, it is necessary to have an economical (compact) representation of state sets. One such candidate is called the *characteristic function*, or *predicate*, of a set. However, as further explained in Chapter 4 of this book,

[2] TCT is software for RW supervisory control, available at `http://www.control.utoronto.ca/DES`.

only models with structure can, generally, have predicates that are appreciably simpler than the state sets represented.

Structure makes a crucial difference in our synthesis. That is why we need a modelling tool like Statecharts [Har87], which offer a compact representation of hierarchy (vertical) and concurrency (horizontal) structure in finite state machines (FSM). Here the system state space is structured top-down into successive layers of cartesian products (AND superstates) alternating with disjoint unions (OR superstates). On this basis, Wang [Wan95] introduced *State Tree Structures* (STS) consisting of a hierarchical state space or *State Tree* (ST), equipped with dynamic modules called *holons* (inspired by [Koe89] and [GHL94]). In [Wan95], however, AND states had to be converted by synchronous product of factors into OR states at a higher level before computations could effectively be carried out. Unfortunately, if the root state is an AND superstate, such naive treatment of AND states may result in collapse of the entire hierarchy and thus leave us with an unstructured flat automaton model to work with. Gohari [GW98] formalized Wang's model in linguistic terms, but [GW98] was similarly restricted to a purely OR state expansion.

Statecharts also underlie the *Asynchronous Hierarchical State Machine* (AHSM) model of [BH93], but there shared events among AND components were ruled out. An efficient algorithm was given to solve reachability problems by exploiting the hierarchical structure and asynchronism. We consider the asynchronism restrictive and will introduce a more generalized statecharts setting.

Marchand *et al.* [MG02] recently introduced another simplified version of statecharts, the *Hierarchical Finite State Machines* (HFSM). But HFSM rule out shared events, are apparently restricted to be OR structures at the topmost level, and allow only specifications of special forbidden state type.

The related works just cited attempt to avoid either AND structure or shared events among components. By contrast, in our STS model, developed directly from Wang's STS model, we treat both AND and OR states on a more equal footing, and in AND states allow shared events among the factors.

To design controllers for this more generalized STS framework, we exploit a powerful computational representation of predicates: the *Binary Decision Diagram* (BDD) [Bry86]. A BDD is a rooted directed acyclic graph (DAG) that is used to represent a decision tree for a boolean function $f : \{0,1\}^n \rightarrow \{0,1\}$. In many cases, a BDD is much smaller than the explicit truth table representation of the boolean function. Fig. 1.2 shows a BDD representation of the boolean function $f(x, y) := (x \vee y)$. Given a particular variable assignment, the value of the function is decided in the graph by starting from the root node (labelled by x in Fig. 1.2), at each node branching according to the value of the variable labelling that node. If this traversal leads to the terminal node labelled by the value 1, then $f(x, y)$ is true under this variable assignment. For example, $f(x, y)$ returns value 1 under the variable assignment $\{x \Leftarrow 1, y \Leftarrow 0\}$, as the traversal leads to the terminal node 1. Notice that the BDD in Fig. 1.2 can be looked at as the characteristic function (or symbol) of the

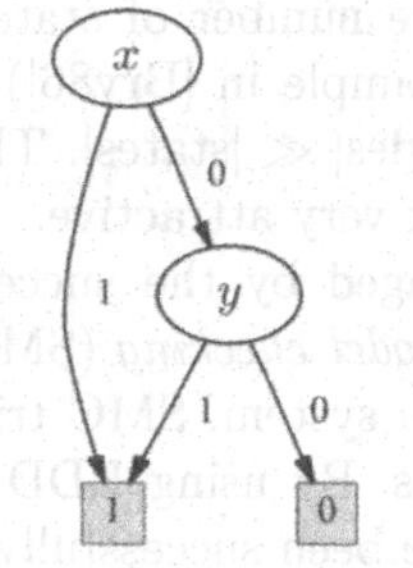

Fig. 1.2. A BDD representation of the boolean function $f := (x \vee y)$

following product set

$$\{(x, y) | x, y \in \{0, 1\} \ \& \ (x \vee y) = 1\},$$

which has 3 elements $(0, 1), (1, 0), (1, 1)$. As a complex DES usually consists of a number of concurrent components and its state space is given as a product set, it is attractive to introduce BDD to represent such a state space in the computer.

A BDD representation is appealing because, for a fixed order of arguments (variables), it is a canonical form. It makes such tasks as testing for equivalence, satisfiability or tautology very simple. Let u_1, u_2 be two BDD and denote by $|u|$ the number of nodes of u. Table 1.1 lists the computational complexity of a few useful BDD operations (from the BDD package *buddy 2.0* [And97]).[3]

Table 1.1. Computational complexity of some BDD operations

Operation	Computational complexity	Description
$u_1 \wedge u_2$	$O(\|u_1\| \times \|u_2\|)$	Conjunction of u_1 and u_2
$u_1 \vee u_2$	$O(\|u_1\| \times \|u_2\|)$	Disjunction of u_1 and u_2
$\neg u_1$	$O(\|u_1\|)$	Negation of u_1
$u_1 \equiv u_2?$	$O(1)$	Test if u_1 and u_2 are equivalent

After representing predicates by BDD, the computational complexity of our synthesis is no longer polynomial in the model's state size, but in the number of BDD nodes in use. In the worst case, the number of BDD nodes (|nodes|) can grow exponentially in the number of variables (blow-up) and

[3] We choose to list these important operators because (i) the set intersection, union, and complement can be *symbolically* computed by the BDD conjunction, disjunction and negation, respectively; (ii) the logic equivalence is necessary in the fixpoint computation of chapters 3 and 4.

therefore is comparable to the number of states (|states|) of the set it represents (see the multiplier example in [Bry86]). However, in many practical complex systems we have $|\text{nodes}| \ll |\text{states}|$. This makes the introduction of BDD into our STS framework very attractive.

Our work is also encouraged by the successful application of BDD in the research area *symbolic model checking* (SMC), initiated by Clarke *et al.* [BCM+92]. Given a transition system, SMC tries to verify if the model satisfies given logic specifications. By using BDD, several extremely large systems [CGH+93, CCM95] have been successfully verified by this technique. It is claimed in [BCC98] that SMC is capable of verifying systems having as many as 10^{120} states. In [BLA+99], the reachability of an example with 10^{400} states is successfully checked by exploiting certain hierarchical structure.

Considering the fact that SMC also faces NP-hard computation problems, we are confident that similar success can be achieved in our STS framework with the help of BDD. In [HWT92, BHG+93], Hoffmann *et al.* presented a BDD-based symbolic computation of RW supervisor for an example of 10^6 states. Its computation capability is limited by a less efficient encoding of the system's transition graph. This makes their approach not so promising because TCT can also handle the example. The breakthrough finally came when Zhang [ZW01] developed S(mart)TCT based on the *Integer Decision Diagram* [Gun97] (IDD), a BDD-like decision diagram. STCT could handle systems with more than 10^{20} states. However, Zhang's work only focused on synchronous product structure and he did not present a detailed theory to support his computation. In this book, we theoretically develop a complete symbolic approach for the optimal nonblocking supervisor design of STS.

There are other types of symbolic computation of a supervisor. One attempt was made by Minhas [Min02, MW03], where a symbolic approach for online supervision of DES was presented to ensure controllability and deadlock-freeness. However, deadlock-freeness is a much easier property to verify than nonblocking, as it is 'local' rather than 'global'. In this book, our interest will be placed mostly on ensuring nonblocking.

In summary, after more than twenty years of effort, we still face two major challenges:

1. How to model and control complex systems?
2. How to make the controller more transparent, especially to users who are not specialists in DES?

In this book, we will propose our answer to both questions.

1.3 Contribution and Outline

The central theoretical contribution of this book can be summarized as: a formalized STS model, the symbolic synthesis of its optimal nonblocking supervisor, and a neat control implementation. Any optimal nonblocking supervisory control problem posed in the original RW framework can be treated in our STS framework without *any* restriction; and our STS framework is readily adaptable to structured complex systems having state size more than 10^{20}.

The rest of this book is organized as follows:

- In chapter 2, a formalized STS model is developed. Several important concepts are introduced. Among them are
  - *state tree* that assigns structure to the state space,
  - *sub-state-tree* that plays the same role as state in an automaton, [4]
  - *holon* that defines the local transition graph.

  The state tree is called a *symbol* of the state space. In the terminology of set theory [FBHL84], the state tree assigns an *intensional* description to the state space (which is a set). [5] This is a description of how the state space is going to be presented or computed. From the state tree, we can find all members of the state space, its *extensional* description. [6] The intensional definition of the state space is essential in our framework because we cannot afford the computer memory to explicitly store all members of the state space of a complex system.

  The dynamics of the entire system is given by the forward transition function Δ, defined over the set of all sub-state-trees, parallel to the δ function defined over the state set in an unstructured automaton. This setup is important because we cannot afford to store the whole transition graph for complex systems. But we can store the dynamics of each holon in the STS and write an algorithm to *compute* Δ. A transition graph is in fact a *set* of transitions. So in terms of set theory, our construction of transition sets by the Δ function is also *intensional*, compared with the *extensional* definition of the transition sets by the δ function in an automaton, i.e., the *explicit* enumeration of all members of the transition sets. We will see that our intensional Δ function has significant computational advantage over the extensional δ function.
- In chapter 3, a new approach to optimal nonblocking supervisory control design is obtained. An innovation in our setting is that reachability does not play as significant a role as in the original RW framework (see chapter 7 of [Won04]), as no computation is required to ensure reachability in our setting. However, given a control problem, both settings generate the *same controlled behavior*, while our setting is computationally more efficient.
- Chapter 4 is the core of the book. Here we demonstrate the symbolic representation and synthesis of STS. First, the control problem for STS is

[4] However, unlike state, sub-state-tree still *retains* structural information.

[5] In set theory, the *extensional* nature of a set means that the set is completely determined by its members. For example, to decide if two sets are equal using extensionality, one needs to compare all members of both sets. This extensional notion of set is conceptually simpler and clearer, but can be computationally more expensive than the intensional description of a set.

[6] We will explain how to do this in Chapter 2, by a concept called *basic sub-state-tree*.

formulated. Second, by introducing the recursive function Θ, the hierarchical state space and transition structures can be easily encoded. We do not encode the entire transition graph by a single complex predicate as in [BCC98], but instead encode it by a set of simpler predicates. This effectively avoids the BDD nodes blow-up problem at this stage. [7] Third, by taking advantage of the rich structure of our STS model, the algorithm given in chapter 3 is adapted to a recursive symbolic algorithm. Auxiliary techniques, e.g., clusters and inference, are also introduced to make the computation even more efficient. Finally, the reader will (we hope) find that the control implementation in our framework is elegant and transparent. Tutorial examples are given to demonstrate how simple our controllers can be.

- Chapter 5 explains how to model and control a benchmark example, the Production Cell, using the STS methodology. The most important advantage of STS demonstrated here is that it can provide users with an integrated view of the control problem. So the STS framework should be more easily accepted by non-specialists in DES.
  This example has on order 10^8 states and our BDD-based program finished the synthesis in less than 5 seconds on a personal computer with 1G Hz Athlon CPU and 256 MB RAM.
- In chapter 6 another case study is presented, the AIP. This example also demonstrates the modelling and synthesis power of our STS framework. It is easy to understand the dynamics of AIP as well as the specifications written by predicates.
  The synthesis result is better than that from any other available synthesis tools in terms of time and space: with state size 10^{24}, our BDD-based program finished the synthesis in less than 20 seconds on a PC with 1G Hz Athlon CPU and 256 MB RAM.
- Finally, chapter 7 presents conclusions and some future research directions. An intuitive discussion of computational complexity is also presented here.

In this book, chapters 2, 3, 4 are the main conceptual chapters. It is suggested that on first reading, the case studies in chapters 5 and 6 can be skipped.

The book brings together and draws freely on the following three research areas: RW framework, statecharts, and symbolic computation using BDD. So it is recommended that serious readers become familiar with the following background material:

- Chapters 3 and 7 of the *Supervisory Control of Discrete-Event Systems* [Won04], available at `http://www.control.utoronto.ca/DES`. These chapters provide the basics of the RW framework.
- the classic paper on statecharts: *Statecharts: A Visual Formalism for Complex Systems* [Har87].

---

[7] In Symbolic Model Checking, one serious problem is that the number of BDD nodes of the overall transition graph can reach millions or more.

- the classic paper on BDD: *Graph-Based Algorithms for Boolean Function Manipulation* [Bry86].
- a helpful survey on symbolic model checking (SMC): *Compositional Reasoning in Model Checking* [BCC98].

Acknowledgements

This book is a mildly revised version of the doctoral thesis of the first author. We would like to thank Professor Jan H. van Schuppen for his critical reading of the thesis as External Appraiser. Professor Mireille Broucke made helpful suggestions for the exposition in Chapter 2. Raoguang Song's patient reading of the thesis has also resulted in certain clarifications. Of course, any errors and obscurities that remain are the authors' sole responsibility.

2

State Tree Structures: Basics

Complex DES are often equipped with hierarchy and concurrency structure. It is often a challenge to incorporate these features into a *compact* and *natural* model. Having this in mind, Bing Wang in 1995 introduced State Tree Structures (STS) having both hierarchy and concurrency[Wan95], and showed them to be a powerful modelling tool. For example, both automaton and synchronous product models are special STS. However, Wang's STS framework was given largely intuitively and the global behavior of STS was left unclear.

P. Gohari [Goh98] developed a linguistic version of Wang's STS framework in 1998, using *structured formal language*. However, this still did not formalize the important concept of parallel (AND) decomposition, or concurrency, in Wang's STS framework.

We introduce in this chapter a complete formalization of Wang's STS. In section 2.1, we define the *State Tree*, a hierarchical state space featuring horizontal (concurrency) and vertical (hierarchy) structure; in section 2.2, we define the *Holon*, a local transition graph assigned to a superstate on the State Tree; in section 2.3, we define the *State Tree Structure*, namely the state tree with its associated set of holons, and we specify completely its global behavior.

Modelling complex DES by STS is an art. Designers are free to bring in levels (hierarchy) and concurrency as they see fit. There is no precise answer as to how many levels a good STS model should have and what types of DES can be best handled in this framework. In this chapter, we assume that the STS modelling decisions have already been made, and define the basic concepts on which the STS model rests. In particular, the STS model is not refined to improve or optimize complexity. However, the case studies in Chapters 5 and 6 will provide some modelling heuristics (cf. section 5.2.8).

2.1 State Tree

The state space of a complex DES may have billions of elements, so we cannot afford to store the entire state space *explicitly* by listing all of its elements in the computer. In this section, the notion of a *state tree* is introduced to economically represent a state space.

To start with, we introduce a concept to describe state aggregation and layering.

Definition 2.1 [AND/OR Superstate [1]]

Let X be a finite collection of sets, called *states*, and let $x \in X$. Suppose $Y = \{x_1, x_2, \ldots, x_n\} (n \geq 1)$ is an element of 2^X, the power set of X, where $x \notin Y$. We call x a *superstate in X expanded by Y* if x can be represented by all of the states in Y, using either expansion described as follows.

1. As illustrated in Figure 2.1.(a), x is the disjoint union of states in Y. Write $x = \dot{\bigcup}_{x_i \in Y} x_i$. We call x an *OR superstate* and each x_i an *OR component* of x. Disjointness means that the semantics of x is the *exclusive-or* (XOR) of $x_i, i = 1, 2, \ldots, n$, namely to be "at" state x the system must be at *exactly one* state of Y.
2. As illustrated in Figure 2.1.(b), x is the cartesian product of states in Y. Write $x = \prod_{x_i \in Y} x_i$, or $x = x_1 \times x_2 \ldots \times x_n$, or simply $x = (x_1, x_2, \ldots, x_n)$. We call x an *AND superstate* and each x_i an *AND component* of x. The semantics of x is then the *and* of $x_i, i = 1, 2, \ldots, n$, i.e., to be at state x the system must be at *all* states of Y simultaneously.

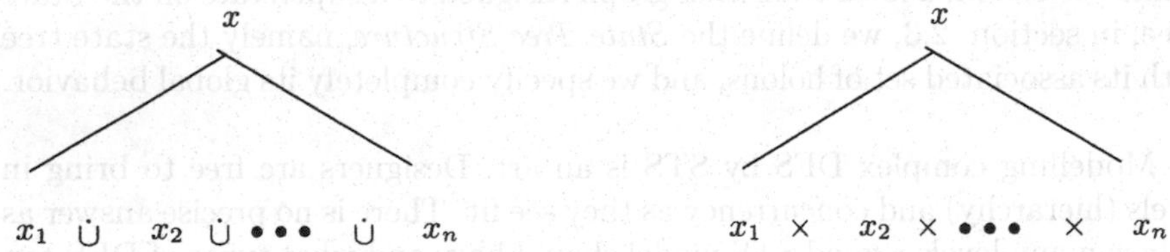

(a) OR superstate (b) AND superstate

Fig. 2.1. State Expansion

In both expansions, we call x the *parent* of x_i and x_i a *child* of x; or say x is *expanded* by the state set Y and the state set Y *expands* x.

◊

[1] The notion of AND/OR is from Statecharts [Har87] and should not be confused with the boolean conjunction/disjunction.

Remarks

1. In our framework, the fundamental concept of AND/OR superstates provides us with means to organize the state set X in a level-wise fashion.
2. The state set X with superstates is also called a *structured state set.* The states other than superstates in X are *simple states,* which do not have any nonempty state sets to further expand them.

The concept of *superstate* is fundamental and we introduce two new functions to describe it more precisely. Let X be a structured state set. Define

$$\mathcal{T} : X \longrightarrow \{and, or, simple\}$$

as the *type function* , such that for all $x \in X$,

$$\mathcal{T}(x) := \begin{cases} and, & \text{if } x \text{ is an AND superstate} \\ or, & \text{if } x \text{ is an OR superstate} \\ simple, & \text{otherwise} \end{cases} .$$

Write $\mathcal{T}_R$ for the restriction of $\mathcal{T}$ to a subset $R \subseteq X$. Formally, $\mathcal{T}_R : R \longrightarrow \{and, or, simple\}$ such that

$$\mathcal{T}_R(x) := \mathcal{T}(x), \quad x \in R.$$

Define

$$\mathcal{E} : X \longrightarrow 2^X$$

as the *expansion function* such that, with $x \in X$, $\emptyset \subset Y \subseteq X$ and $x \notin Y$,

$$\mathcal{E}(x) := \begin{cases} Y, \text{ if } \mathcal{T}(x) \in \{and, or\} \\ \emptyset, \text{ if } \mathcal{T}(x) = simple \end{cases}, \quad Y \text{ as above,}$$

where $\emptyset$ is the empty state set. Write $\mathcal{E}_R$ for the restriction of $\mathcal{E}$ to a subset $R \subseteq X$. Formally, $\mathcal{E}_R : R \longrightarrow 2^R$ such that, for $x \in R$,

$$\mathcal{E}_R(x) := \mathcal{E}(x) \cap R.$$

For example, if x is an AND superstate expanded by the set Y, then $\mathcal{T}(x) = and$ and $\mathcal{E}(x) = Y$. One special case is that $X = \{x\}$, i.e., X has a single element, implies that $\mathcal{T}(x) = simple$ and $\mathcal{E}(x) = \emptyset$, because the set expanding a superstate must be nonempty and must not include x itself.

It is convenient to extend $\mathcal{E}$ into a sequence of functions $\{\hat{\mathcal{E}}^n\}, n = 1, 2, \ldots$. Let X be a structured state set and $x \in X$. Define $\hat{\mathcal{E}}^n : X \longrightarrow 2^X$ by induction:

$$\hat{\mathcal{E}}^1(x) := \mathcal{E}(x) \cup \{x\},$$
$$\hat{\mathcal{E}}^n(x) := \bigcup_{y \in \hat{\mathcal{E}}^{n-1}(x)} \hat{\mathcal{E}}^1(y).$$

Obviously, for all x, $\hat{\mathcal{E}}^n(x) \subseteq \hat{\mathcal{E}}^{n+1}(x)$, i.e., the series is monotone, in the sense of inclusion. If X is finite, there must exist a function that is the limit of the series. We call it the *reflexive and transitive closure of* $\mathcal{E}$, and denote it by $\hat{\mathcal{E}}^*$. Formally, $\hat{\mathcal{E}}^* : X \longrightarrow 2^X$ is defined by

$$\hat{\mathcal{E}}^*(x) := \lim_{n\to\infty} \hat{\mathcal{E}}^n(x).$$

For simplicity, we omit the ˆ and write $\mathcal{E}^*$ instead of $\hat{\mathcal{E}}^*$. $\mathcal{E}^*$ can be extended to $\mathcal{E}^* : 2^X \longrightarrow 2^X$ because for any $Y \subseteq X$, we can define

$$\mathcal{E}^*(Y) := \bigcup_{x\in Y} \mathcal{E}^*(x).$$

Lemma 2.1

1. $\mathcal{E}^*$ is monotone, i.e., $(\forall Y, Z \subseteq X) Y \subseteq Z \Rightarrow \mathcal{E}^*(Y) \subseteq \mathcal{E}^*(Z)$.
2. $(\forall Y \subseteq X)\mathcal{E}^*(Y) = \mathcal{E}^*(\mathcal{E}^*(Y))$.
3. $(\forall x, y \in X) y \in \mathcal{E}^*(x) \Leftrightarrow \mathcal{E}^*(y) \subseteq \mathcal{E}^*(x)$.

Proof. 1 and 2 are directly from the definition. We just need to prove 3.

1. ($\Leftarrow$) automatic as $y \in \mathcal{E}^*(y)$.
2. ($\Rightarrow$) $y \in \mathcal{E}^*(x) \Rightarrow \{y\} \subseteq \mathcal{E}^*(x)$. By item 1 of this lemma, $\mathcal{E}^*(\{y\}) \subseteq \mathcal{E}^*(\mathcal{E}^*(x))$. Also $\mathcal{E}^*(\{y\}) = \mathcal{E}^*(y)$, $\mathcal{E}^*(\mathcal{E}^*(x)) = \mathcal{E}^*(x)$. So $\mathcal{E}^*(y) \subseteq \mathcal{E}^*(x)$ as required.

Let $\mathcal{E}^+ : X \longrightarrow 2^X$ be the *unfolding* of $\mathcal{E}$, defined by

$$\mathcal{E}^+(x) := \mathcal{E}^*(x) - \{x\}, \forall x \in X.$$

For convenience, if $y \in \mathcal{E}^+(x)$, we call y a *descendant* of x and x an *ancestor* of y.

Now we can define the *state tree* by recursion.

Definition 2.2 [State Tree] Consider a 4-tuple $(X, x_o, \mathcal{T}, \mathcal{E})$, where

- X is a finite structured state set with $X = \mathcal{E}^*(x_o)$,
- $x_o \in X$ is a special state called the *root state*,
- $\mathcal{T} : X \longrightarrow \{and, or, simple\}$ is the type function,
- $\mathcal{E} : X \longrightarrow 2^X$ is the expansion function.

The 4-tuple $\mathbf{ST} = (X, x_o, \mathcal{T}, \mathcal{E})$ is a *state tree* if

1. (terminal case) $X = \{x_o\}$, or
2. (recursive case) $(\forall y \in \mathcal{E}(x_o))\mathbf{ST}^y = (\mathcal{E}^*(y), y, \mathcal{T}_{\mathcal{E}^*(y)}, \mathcal{E}_{\mathcal{E}^*(y)})$ is also a state tree, where

$$(\forall y, y' \in \mathcal{E}(x_o))(y \neq y' \Rightarrow \mathcal{E}^*(y) \cap \mathcal{E}^*(y') = \emptyset)$$

and

$$\dot{\bigcup}_{y\in\mathcal{E}(x_o)} \mathcal{E}^*(y) = \mathcal{E}^+(x_o).$$

That is, the family of subsets $\{\mathcal{E}^*(y) | y \in \mathcal{E}(x_o)\}$ partitions $\mathcal{E}^+(x_o)$. Say $\mathbf{ST}^y$ is a *child state tree of* x_o in $\mathbf{ST}$, rooted by y.

A *well-formed* state tree must also satisfy

$$(\forall x, y \in X)\mathcal{T}(x) = and \ \& \ y \in \mathcal{E}(x) \Rightarrow \mathcal{T}(y) \in \{or, and\}.$$

That is, no AND components can be simple states. [2] In general, AND components must be superstates, but OR components can be any of the three types of states. We assume that all state trees are well-formed.

For convenience, say **ST** is an *empty state tree* if $X = \emptyset$. Denote it by $\mathbf{ST} = \emptyset$.

◊

Remarks

1. A state tree **ST**, shown in Figure 2.2, provides a general idea of what a state tree looks like. In Figure 2.2, **ST** is well-formed because the children, x_{11} and x_{12}, of the AND superstate x_1 are both OR superstates. In Figure 2.3, we see how the definition works. In **ST**, there are 3 child state trees of x_o, according to x_o's child states. Each child of the root state x_o is the root state of a child state tree. For example, x_1 is the root of the state tree $\mathbf{ST}^{x_1}$, where the children of x_1, namely x_{11} and x_{12}, are both root states of their own child state trees, $\mathbf{ST}^{x_{11}}, \mathbf{ST}^{x_{12}}$, respectively.
2. The above example also demonstrates that, in the recursive case of the definition, $(\forall y, y' \in \mathcal{E}(x_o))(y \neq y' \Rightarrow \mathcal{E}^*(y) \cap \mathcal{E}^*(y') = \emptyset)$ and $\dot{\bigcup}_{y \in \mathcal{E}(x_o)} \mathcal{E}^*(y) = \mathcal{E}^+(x_o)$. For example, $\mathcal{E}(x_o) = \{x_1, x_2, x_3\}$, and

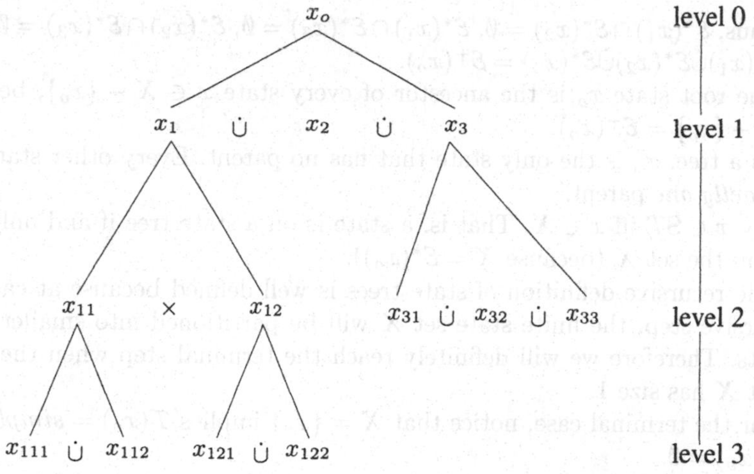

Fig. 2.2. A well-formed State Tree **ST**

[2] This restriction on a state tree is not necessary, but is convenient. Please refer to the following remark (item 9) for explanation.

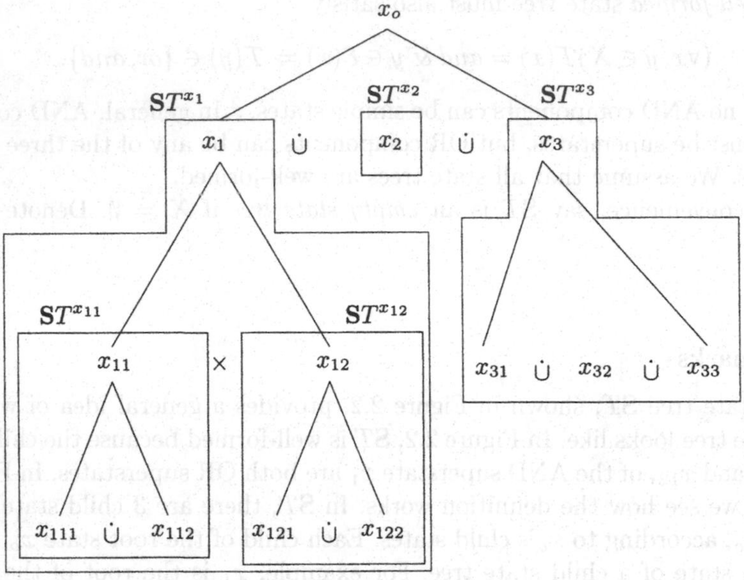

Fig. 2.3. Recursive definition of State Tree **ST**

$$\begin{aligned}
\mathcal{E}^*(x_1) &= \{x_1, x_{11}, x_{111}, x_{112}, x_{12}, x_{121}, x_{122}\},\\
\mathcal{E}^*(x_2) &= \{x_2\},\\
\mathcal{E}^*(x_3) &= \{x_3, x_{31}, x_{32}, x_{33}\}.
\end{aligned}$$

Thus, $\mathcal{E}^*(x_1)\cap\mathcal{E}^*(x_2) = \emptyset$, $\mathcal{E}^*(x_1)\cap\mathcal{E}^*(x_3) = \emptyset$, $\mathcal{E}^*(x_2)\cap\mathcal{E}^*(x_3) = \emptyset$, and $\mathcal{E}^*(x_1)\dot\cup\mathcal{E}^*(x_2)\dot\cup\mathcal{E}^*(x_3) = \mathcal{E}^+(x_o)$.

3. The root state x_o is the ancestor of every state $x \in X - \{x_o\}$, because $X - \{x_o\} = \mathcal{E}^+(x_o)$.
4. As a tree, x_o is the only state that has no parent. Every other state has *exactly one* parent.
5. Say $x \in \mathbf{ST}$ iff $x \in X$. That is, a state is on a state tree if and only if it is in the set X (because $X = \mathcal{E}^*(x_o)$).
6. The recursive definition of state trees is well defined because at each recursive step, the finite state set X will be partitioned into smaller state sets. Therefore we will definitely reach the terminal step when the state set X has size 1.
7. For the terminal case, notice that $X = \{x_o\}$ implies $\mathcal{T}(x_o) = \textit{simple}$ and $\mathcal{E}(x_o) = \emptyset$.
8. For the recursive case, notice that $x_o \notin \mathcal{E}(x_o)$ by the definition of $\mathcal{E}$, which means the state tree is *acyclic*. Also the set $(X - \{x_o\}) = \mathcal{E}^+(x_o)$ is partitioned into disjoint sets: $\mathcal{E}^*(x_1), \mathcal{E}^*(x_2), \ldots, \mathcal{E}^*(x_n)$, which rules out the following scenarios, shown in Figure 2.4. In case (a), state x_3 is not linked to the tree. In case (b), x_3 has two different parents.

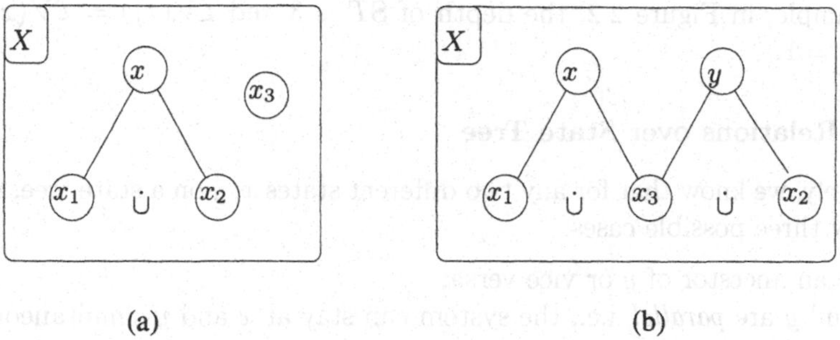

Fig. 2.4. Illegal scenarios on a state tree

9. The objective of introducing the concept of state tree is to build our STS model. Here we require a state tree to be *well-formed.* This does not mean that we cannot build our STS model on an arbitrary state tree; but rather, we do not have to. We explain this as follows. A well-formed state tree rules out only one case: some AND components are simple states. For in that case, there would be no way to assign transitions to those simple states [3]. So we can cut them from the state tree because they are redundant in terms of describing the system behavior (provide no additional information).
10. A state tree with $X = \{x_o\}$ is trivially well-formed (having no superstates).

In a well-formed state tree, no simple states can be AND components. This also guarantees the following important property.

Lemma 2.2 Let $\mathbf{ST} = (X, x_o, \mathcal{T}, \mathcal{E})$ be a state tree. Assume the size of X is greater than 1. Then all *leaf states*, the states without children, must be OR components.

Proof. Let y be a leaf state. Then $\mathcal{T}(y) = simple$ as all superstates have children. Let $x \in X$ and $y \in \mathcal{E}(x)$. If $\mathcal{T}(x) = and$, then $\mathcal{T}(y) \in \{or, and\}$ by the definition of well-formed state tree, a contradiction. Thus $\mathcal{T}(x) = or$ as required.

Definition 2.3 [Level, Depth] Let $\mathbf{ST} = (X, x_o, \mathcal{T}, \mathcal{E})$ be a state tree. Define the *level function* $\mathcal{LV} : X \longrightarrow N$ such that for all $x \in X$

$$\mathcal{LV}(x) = \begin{cases} 0, & \text{if } x = x_o \\ \mathcal{LV}(z) + 1, \text{ where } x \in \mathcal{E}(z), & \text{if } x \neq x_o \end{cases}.$$

There must exist a unique maximal level number in $\mathbf{ST}$. Call it the *depth* of $\mathbf{ST}$.

◇

[3] The dynamics of STS will be described in detail in the two sections to follow. For now, the reader is requested to accept this statement on faith. The reason will become clear after reading section 2.3 up to page 38.

For example, in Figure 2.2, the depth of **ST** is 3 and $\mathcal{LV}(x_1) = \mathcal{LV}(x_2) = \mathcal{LV}(x_3) = 1$.

2.1.1 Relations over State Tree

Intuitively, we know that for any two different states x, y on a state tree, there are only three possible cases:

1. x is an ancestor of y or vice versa;
2. x and y are *parallel*, i.e., the system can stay at x and y simultaneously;
3. x and y are *exclusive*, i.e., the system cannot stay at x and y at the same time.

We will define the above three relations one by one in this subsection.

Definition 2.4 [$\leq$] Let $\mathbf{ST} = (X, x_o, \mathcal{T}, \mathcal{E})$ be a state tree. Let $x, y \in X$ be any two states in $\mathbf{ST}$. Define

$$x \leq y \text{ iff } y \in \mathcal{E}^*(x).$$

◇

Intuitively, $x \leq y$ if either $x = y$, or x is an ancestor of y.

Lemma 2.3 Let $\mathbf{ST} = (X, x_o, \mathcal{T}, \mathcal{E})$ be a state tree. Then

$\leq$ is a *partial order (p.o.)* on X.

Proof. We need to prove the following 3 assertions.

1. (Reflexive). Show that $(\forall x \in X) x \leq x$.
   This is automatic from $x \in \mathcal{E}^*(x)$.
2. (Transitive). Show that $(\forall x, y, z \in X) x \leq y \,\&\, y \leq z \Rightarrow x \leq z$. We have $x \leq y \Rightarrow y \in \mathcal{E}^*(x)$. By item 3 of Lemma 2.1, $\mathcal{E}^*(y) \subseteq \mathcal{E}^*(x)$. Also $y \leq z \Rightarrow z \in \mathcal{E}^*(y)$. Then $z \in \mathcal{E}^*(x)$ as required.
3. (Antisymmetric). Show that $(\forall x, y \in X) x \leq y \,\&\, y \leq x \Rightarrow x = y$.
   If X has a single element, then $x = y$ automatically. Otherwise we prove it by contradiction. From $x \leq y$, we have $\mathcal{E}^*(y) \subseteq \mathcal{E}^*(x)$ by Lemma 2.1. Assume $x \neq y$. Then from the recursive step of the definition of state tree and $\mathcal{E}^*(y) \subseteq \mathcal{E}^*(x)$, $\mathcal{E}^*(y)$ must also be a subset of $\mathcal{E}^*(x) - \{x\} = \mathcal{E}^+(x)$. However, $y \leq x \Rightarrow x \in \mathcal{E}^*(y)$. Therefore, $x \in \mathcal{E}^+(x)$, a contradiction.

Finally we can say $\leq$ is a partial order on X.

After defining $\leq$ on X, we call the set X a *poset*. In a poset X, in general, not every two arbitrary elements are comparable. If x and y are not comparable, we write $x <> y$. If every two elements of X are comparable, X is *totally ordered*. The root state x_o, the ancestor of all other states in X, is the *bottom element* of X.

It is useful to have some new notation. Write $x < y$ if $x \leq y$ and $x \neq y$, i.e., x is an ancestor of y. Write $x \not< y$ if $y \leq x$ or $x <> y$.

Lemma 2.4 Let $\mathbf{S}\mathcal{T} = (X, x_o, \mathcal{T}, \mathcal{E})$ be a state tree. Let $A_x = \{a | a < x\}$ be the ancestor set of x. Then A_x is totally ordered.

Proof. First we show how to compute A_x on a state tree.

Let $B = \emptyset$.

If $x = x_o$, we terminate with $A_x = B = \emptyset$, because no states can be the ancestors of x_o.

Assume $x \neq x_o$. Then add x_o to B as x_o is the ancestor of any other state in X. If $x \in \mathcal{E}(x_o)$, we terminate with $A_x = B$. If not, from the definition of state tree, there must exist *exactly one* child state tree of x_o, say $\mathbf{S}\mathcal{T}^{x_1} = (\mathcal{E}^*(x_1), x_1, \mathcal{T}_{\mathcal{E}^*(x_1)}, \mathcal{E}_{\mathcal{E}^*(x_1)})$, such that $x \in \mathcal{E}^+(x_1)$. All remaining ancestors of x must be in $\mathbf{S}\mathcal{T}^{x_1}$, because $(\forall y \in \mathcal{E}(x_o), y \neq x_1) x \notin \mathcal{E}^*(y)$. Repeat the same process on the child state tree $\mathbf{S}\mathcal{T}^{x_1}$ as on $\mathbf{S}\mathcal{T}$ until $x \in \mathcal{E}(x_n)$, where x_n is the root of a child state tree $\mathbf{S}\mathcal{T}^{x_n}$.

Now we have in general $A_x = B = \{x_o, x_1, \ldots, x_n\}$. If A_x is nonempty, it is totally ordered because

$$\mathcal{E}^*(x_o) \supset \mathcal{E}^*(x_1) \supset \ldots \supset \mathcal{E}^*(x_n).$$

If A_x is empty, it is trivially totally ordered.

For example, in Figure 2.2 on page 15, $A_{x_{111}} = \{x_o, x_1, x_{11}\}$ is totally ordered.

Definition 2.5 [Nearest Common Ancestor [HT84]] In a state tree, state c is the *nearest common ancestor (NCA)* of two states a, b if c satisfies the following two conditions:

1. c is an ancestor of a and b, i.e., $c < a$ and $c < b$;
2. no descendant of c is an ancestor of both a and b, i.e.,

$$(\forall x \in \mathcal{E}^+(c)) x \not< a \text{ or } x \not< b.$$

◇

Lemma 2.5 Let $\mathbf{S}\mathcal{T} = (X, x_o, \mathcal{T}, \mathcal{E})$ be a state tree. Let $x, y \in X$ and $x <> y$. Then there exists a unique NCA of x, y.

Proof. $x <> y$ implies $x, y \neq x_o$, the root state. So $x_o \in A_x$, the ancestor set of x and $x_o \in A_y$, the ancestor set of y. This means $A_x \cap A_y \neq \emptyset$. Because A_x, A_y are both finite and totally ordered, so is $A_x \cap A_y$ and there must exist a unique top element z in $A_x \cap A_y$, which is the NCA of x, y.

An algorithm is given in [HT84] to compute NCA. Its time computational complexity is $O(n)$, where n is the number of states on the tree.

Definition 2.6 [$|$, $\oplus$] Let $\mathbf{S}\mathcal{T} = (X, x_o, \mathcal{T}, \mathcal{E})$ be a state tree. Let $x, y \in X$ and $x <> y$.

Define

$$x|y \text{ iff the NCA of } x \text{ and } y \text{ is an AND superstate.}$$

Call x and y *parallel.* Let $A, B \subset X$. Say $A|B$ if

$$(\forall x \in A, y \in B) x|y.$$

Define

$$x \oplus y \text{ iff the NCA of } x \text{ and } y \text{ is an OR superstate.}$$

Call x and y *exclusive.* Say $A \oplus B$ if

$$(\forall x \in A, y \in B) x \oplus y.$$

◇

The interpretation of $x|y$ is that, in a run of the system, the system state can visit x and y simultaneously(concurrently), while $x \oplus y$ if the system state can visit either x or y but not both simultaneously. For example, as illustrated in Figure 2.5, $x_{11}|x_{12}$ and $\{x_{11}, x_{111}, x_{112}\}|\{x_{12}, x_{121}\}$; $x_1 \oplus x_{31}$ and $\{x_1, x_{11}, x_{111}\} \oplus \{x_2, x_3, x_{31}\}$.

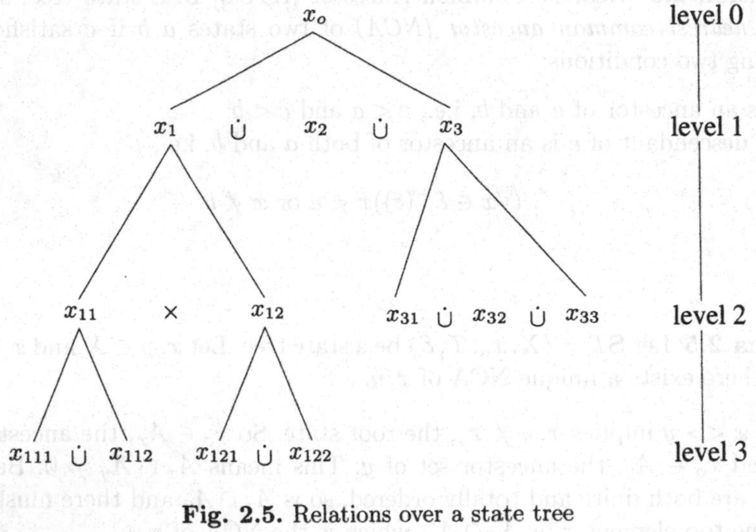

Fig. 2.5. Relations over a state tree

Lemma 2.6 Let $\mathbf{ST} = (X, x_o, \mathcal{T}, \mathcal{E})$ be a state tree. Let $x, y \in X$. Then x, y satisfy exactly one of the following

1. $x \leq y$ or $y \leq x$,
2. $x|y$,
3. $x \oplus y$.

Proof. If x and y are comparable, $x \leq y$ or $y \leq x$. If $x <> y$, there exists a unique NCA of x and y by Lemma 2.5. So we have either $x|y$ or $x \oplus y$, as the NCA of x, y must be either an AND superstate or an OR superstate.

2.1.2 Sub-State-Tree

Definition 2.7 [sub-state-tree] Let $\mathbf{ST} = (X, x_o, \mathcal{T}, \mathcal{E})$ be a state tree. Let $Y \subseteq X$. $\mathbf{subST} = (Y, x_o, \mathcal{T}', \mathcal{E}')$ is a *sub-state-tree* of $\mathbf{ST}$ if $\mathbf{subST}$ is a well-formed state tree, with $\mathcal{T}' : Y \longrightarrow \{and, or, simple\}$ and $\mathcal{E}' : Y \longrightarrow 2^Y$ defined by

$$\mathcal{T}'(y) := \mathcal{T}(y),$$

$$\mathcal{E}'(y) := \begin{cases} \mathcal{E}(y), & \text{if } \mathcal{T}'(y) = and \\ Z, \text{ where } \emptyset \subset Z \subseteq \mathcal{E}(y), & \text{if } \mathcal{T}'(y) = or \\ \emptyset, & \text{if } \mathcal{T}'(y) = simple \end{cases},$$

for all $y \in Y$. [4] Trivially, the empty state tree $\emptyset$ and $\mathbf{ST}$ itself are sub-state-trees of $\mathbf{ST}$. A *proper* sub-state-tree of $\mathbf{ST}$ is one with $Y \subset X$.

For convenience, define

$$\mathcal{ST}(\mathbf{ST}) := \{\mathbf{subST} | \mathbf{subST} \text{ is a sub-state-tree of } \mathbf{ST}\}$$

as the set of all sub-state-trees of $\mathbf{ST}$.

◊

Remarks

1. The state tree $\mathbf{ST}_1$ is a sub-state-tree of $\mathbf{ST}$, shown in Figure 2.6. However, $\mathbf{ST}_2$ is not a sub-state-tree of $\mathbf{ST}$ as x_{12} is an OR superstate but has no children on $\mathbf{ST}_2$, which is not allowed on a sub-state-tree (violates $\emptyset \subset Z$ in the definition).
2. Notice that $\mathbf{subST}$ is still a well-formed state tree. So we can also define by inheritance those relations in the last subsection over $\mathbf{subST}$.
3. Notice that $Y = \mathcal{E}'^*(x_o)$ because $\mathbf{subST}$ is a state tree. That is, x_o is the ancestor of all other states in Y. A sub-state-tree of $\mathbf{ST}$ should not be confused with a child state tree of x_o.
4. $\mathcal{T}' = \mathcal{T}_Y$, i.e., $\mathcal{T}'$ is the restriction of $\mathcal{T}$ to Y. To $\mathcal{E}'$, the only difference is that when y is an OR superstate, $\mathcal{E}'(y)$ is a nonempty subset of $\mathcal{E}(y)$. Notice that for all the states in $\mathcal{E}(y) - \mathcal{E}'(y)$, their descendants on $\mathbf{ST}$ are not allowed in Y as $\mathbf{subST}$ is a state tree. Intuitively, cutting off the branches (child state trees) rooted by OR components will result in a sub-state-tree as long as every OR state on the resulting state tree has at least one child (i.e., not too much is cut).

[4] The requirement for $\emptyset \subset Z$ is necessary to preserve the 'type' of y, i.e., in order for y to be OR state, it must have at least one child.

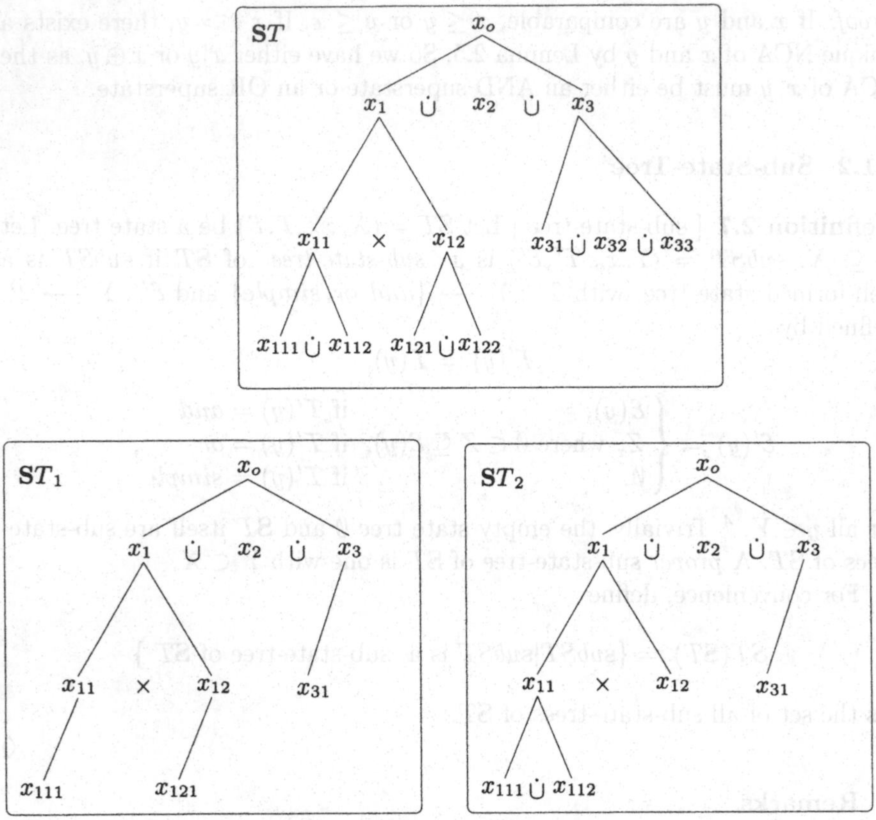

Fig. 2.6. Example: sub-state-trees of **ST**

5. In statecharts, a counterpart concept (of sub-state-tree) called *configuration* is used to define the semantics of statecharts. The reader interested in statecharts is referred to [EGKP97]. We will use sub-state-tree to define the behavior of state tree structures (STS).

Definition 2.8 [$\leq$] Let $\mathbf{ST} = (X, x_o, \mathcal{T}, \mathcal{E})$ be a state tree. Let $\mathbf{ST}_1, \mathbf{ST}_2 \in \mathcal{ST}(\mathbf{ST})$. Define

$$\mathbf{ST}_1 \leq \mathbf{ST}_2 \text{ iff } \mathbf{ST}_1 \in \mathcal{ST}(\mathbf{ST}_2).$$

That is, $\mathbf{ST}_1 \leq \mathbf{ST}_2$ if and only if $\mathbf{ST}_1$ is a sub-state-tree of $\mathbf{ST}_2$. ◊

The following lemma will make it easy to verify the relation $\leq$.

Lemma 2.7 Let $\mathbf{ST} = (X, x_o, \mathcal{T}, \mathcal{E})$ be a state tree. Let $\mathbf{ST}_1 = (X_1, x_o, \mathcal{T}_1, \mathcal{E}_1)$ and $\mathbf{ST}_2 = (X_2, x_o, \mathcal{T}_2, \mathcal{E}_2)$ be two sub-state-trees of $\mathbf{ST}$. Then

$$\mathbf{ST}_1 \leq \mathbf{ST}_2 \text{ iff } X_1 \subseteq X_2.$$

Proof. The statement is obvious when $X_1 = \emptyset$. Otherwise we can prove it as follows.

1. (⇒) Directly from the definition of sub-state-tree.
2. (⇐) We prove it along the state tree. First we know $x_o \in X_1 \subseteq X_2$ for nonempty state trees. Then it is sufficient to show $(\forall x \in X_1)\mathcal{E}_1(x) \subseteq \mathcal{E}_2(x)$ as both $\mathbf{ST}_1$ and $\mathbf{ST}_2$ are state trees. It is obvious that $\mathcal{E}_1(x) = \mathcal{E}_2(x) = \mathcal{E}(x)$ when x is a simple state or an AND superstate. When x is an OR superstate, we prove $\mathcal{E}_1(x) \subseteq \mathcal{E}_2(x)$ by contradiction. Suppose $y \in X_1$ and $y \in \mathcal{E}_1(x) \subseteq \mathcal{E}(x)$ but $y \notin \mathcal{E}_2(x)$. Then $y \in \mathcal{E}(x) - \mathcal{E}_2(x)$, which means y is not on $\mathbf{ST}_2$ because $\mathbf{ST}_2$ is a state tree. That is, $y \notin X_2$, contradicting $X_1 \subseteq X_2$.

Since $\subseteq$ is a partial order over the state power set 2^X, so is $\leq$ over $\mathcal{ST}(\mathbf{ST})$ from Lemma 2.7.

Lemma 2.8 Let $\mathbf{ST} = (X, x_o, \mathcal{T}, \mathcal{E})$ be a state tree. Then $\leq$ is a *partial order (p.o.)* on $\mathcal{ST}(\mathbf{ST})$.

Proof. We need to prove that $\leq$ is reflexive, transitive and antisymmetric. These properties are direct from the definition.

In $\mathcal{ST}(\mathbf{ST})$, the empty state tree $\emptyset$ and $\mathbf{ST}$ are the *bottom element* and *top element*, respectively, just as the empty set $\emptyset$ and X are the bottom element and top element of 2^X.

Proposition 2.1 [meet] [5] In the poset $(\mathcal{ST}(\mathbf{ST}), \leq)$, the *meet* $\mathbf{ST}_1 \wedge \mathbf{ST}_2$ of elements $\mathbf{ST}_1$ and $\mathbf{ST}_2$ always exists, and is given as follows.

1. If either $\mathbf{ST}_1$ or $\mathbf{ST}_2$ is empty, $\mathbf{ST}_1 \wedge \mathbf{ST}_2 = \emptyset$.
2. Now suppose $\mathbf{ST}_1$ and $\mathbf{ST}_2$ are nonempty. Let $\mathbf{ST}_1 = (X_1, x_o, \mathcal{T}_1, \mathcal{E}_1)$ and $\mathbf{ST}_2 = (X_2, x_o, \mathcal{T}_2, \mathcal{E}_2)$. We define $\mathbf{ST}_3 = (X_3, x_o, \mathcal{T}_3, \mathcal{E}_3) = \mathbf{ST}_1 \wedge \mathbf{ST}_2$ by recursion [6]. Notice that we just need to define $\mathcal{E}_3$, as $X_3 = \mathcal{E}_3^*(x_o)$ and $\mathcal{T}_3$ is the restriction of $\mathcal{T}$ to X_3.
   a) Case 1 (terminal case): $\mathcal{T}(x_o) = \mathit{simple}$. Then $\mathcal{E}_3(x_o) = \emptyset$.
   b) Case 2 (recursive case): $\mathcal{T}(x_o) = \mathit{or}$. Then $y \in \mathcal{E}_3(x_o)$ iff
      i. $y \in \mathcal{E}_1(x_o) \cap \mathcal{E}_2(x_o)$, and
      ii. $\mathcal{E}_3(y) = \emptyset \Rightarrow \mathcal{T}_3(y) = \mathit{simple}$. [7]

      $\mathbf{ST}_3 = \emptyset$ iff $\mathcal{E}_3(x_o) = \emptyset$. [8]
   c) Case 3 (recursive case): $\mathcal{T}(x_o) = \mathit{and}$. Then
   $$\mathcal{E}_3(x_o) = \begin{cases} \mathcal{E}_1(x_o) \cap \mathcal{E}_2(x_o) = \mathcal{E}(x_o), & \text{if } (\forall y \in \mathcal{E}(x_o))\mathcal{E}_3(y) \neq \emptyset \\ \emptyset, & \text{otherwise} \end{cases}.$$
   $\mathbf{ST}_3 = \emptyset$ iff $\mathcal{E}_3(x_o) = \emptyset$.

[5] Examples are given to demonstrate the operator after the Proposition.

[6] By recursion we mean defining $\mathcal{E}_3(x_o)$ by $\mathcal{E}_3(y)$, for some $y \in \mathcal{E}_3(x_o)$.

[7] That is, $\mathcal{T}_3(y) \neq \mathit{simple} \Rightarrow \mathcal{E}_3(y) \neq \emptyset$. Precisely, if y is a superstate that belongs to $\mathbf{ST}_3$, then at least some of its children must be on $\mathbf{ST}_3$ too.

[8] If superstate x_o does not have children, x_o cannot be in the resulting state tree, i.e., $\mathbf{ST}_3$ must be the empty state tree.

Proof. Notice that $\mathbf{S}T_3$ is a well-defined sub-state-tree as the type function $\mathcal{T}_3$ is just the restriction of $\mathcal{T}$ to X_3. If X_3 is empty, $\mathbf{S}T_3$ is a sub-state-tree of $\mathbf{S}T$. If X_3 is not empty, then it is also a sub-state-tree of $\mathbf{S}T$ because for all x in X_3,

1. $\mathcal{E}_3(x) = \mathcal{E}(x)$ if $\mathcal{T}_3(x) = and$;
   By the above definition for the case of AND superstates, $\mathcal{E}_3(x) = \mathcal{E}(x)$ or $\mathcal{E}_3(x) = \emptyset$. We prove $\mathcal{E}_3(x) = \mathcal{E}(x)$ by contradiction. Assume $\mathcal{E}_3(x) = \emptyset$. In that case $x \notin X_3$ because
   a) if x is the root state x_o, then $\mathcal{E}_3(x_o) = \emptyset$ implies that $\mathbf{S}T_3$ is an empty state tree from the definition, i.e., $X_3 = \emptyset$;
   b) if $x_o < x$, then the parent of x, say z, must be an OR superstate or an AND superstate. If z is an OR superstate, $\mathcal{E}_3(x) = \emptyset$ and $\mathcal{T}_3(x) = and$ means $x \notin \mathcal{E}_3(z)$; if z is an AND superstate, $\mathcal{E}_3(x) = \emptyset$ means $\mathcal{E}_3(z) = \emptyset$ and therefore $x \notin \mathcal{E}_3(z)$. Thus, $x \notin X_3$.

   This contradicts $x \in X_3$.
2. $\emptyset \subset \mathcal{E}_3(x) \subseteq \mathcal{E}(x)$ if $\mathcal{T}_3(x) = or$.
   Obviously $\mathcal{E}_3(x) \subseteq \mathcal{E}(x)$. $\mathcal{E}_3(x)$ is also nonempty. Otherwise $x \notin X_3$ because
   a) if x is the root state x_o, then $\mathcal{E}_3(x_o) = \emptyset$ implies $X_3 = \emptyset$;
   b) if $x_o < x$, for the parent of x, say z, $x \notin \mathcal{E}_3(z)$ by the definition for the cases of AND and OR superstates, i.e., $x \notin X_3$.

   This contradicts $x \in X_3$.

We still need to show that

1. $\mathbf{S}T_3 \leq \mathbf{S}T_1$ and $\mathbf{S}T_3 \leq \mathbf{S}T_2$.
   It is easy to see that $X_3 \subseteq X_1$, because
   a) $x_o \in X_3 \Rightarrow x_o \in X_1$, and
   b) $(\forall x \in X_3)\mathcal{E}_3(x) \subseteq \mathcal{E}_1(x)$ from the definition.

   Then $\mathcal{E}_3^*(x_o) \subseteq \mathcal{E}_1^*(x_o)$, i.e., $X_3 \subseteq X_1$. Thus $\mathbf{S}T_3 \leq \mathbf{S}T_1$ from Lemma 2.7. Similarly, $\mathbf{S}T_3 \leq \mathbf{S}T_2$.
2. If $\mathbf{S}T_4 \in \mathcal{S}T(\mathbf{S}T)$ and $\mathbf{S}T_4 \leq \mathbf{S}T_1$, $\mathbf{S}T_4 \leq \mathbf{S}T_2$, then $\mathbf{S}T_4 \leq \mathbf{S}T_3$.
   We first show that $(\forall x \in X_4)\mathcal{E}_4(x) \subseteq \mathcal{E}_3(x)$ by recursion.
   a) (terminal case) $\mathcal{T}(x) = simple$. $\mathcal{E}_4(x) = \emptyset$. Thus $\mathcal{E}_4(x) \subseteq \mathcal{E}_3(x)$ as required. This covers the leaf states on $\mathbf{S}T_4$.
   b) (recursive case) We need to show that $(\forall x \in X_4, \mathcal{T}(x) \in \{and, or\})$ $\mathcal{E}_4(x) \subseteq \mathcal{E}_3(x)$ if $(\forall y \in \mathcal{E}_4(x))\mathcal{E}_4(y) \subseteq \mathcal{E}_3(y)$.
      i. $\mathcal{T}(x) = and$. As $\mathbf{S}T_4$ is a sub-state-tree, $\mathcal{E}_4(x) = \mathcal{E}(x)$ and $(\forall y \in \mathcal{E}(x))\mathcal{E}_4(y) \neq \emptyset$ (every superstate y on a sub-state-tree must have at least one child). Then by hypothesis, $(\forall y \in \mathcal{E}(x))\mathcal{E}_3(y) \neq \emptyset$. Thus, $\mathcal{E}_3(x) = \mathcal{E}(x)$ from the definition of meet.
      ii. $\mathcal{T}(x) = or$. Let $y \in \mathcal{E}_4(x)$. As $\mathbf{S}T_4$ is a sub-state-tree of both $\mathbf{S}T_1$ and $\mathbf{S}T_2$, it follows that $y \in \mathcal{E}_1(x) \cap \mathcal{E}_2(x)$ and $\mathcal{E}_4(y) = \emptyset \Rightarrow \mathcal{T}(y) = simple$ (every superstate on a sub-state-tree must have at least one child). In the case that $\mathcal{E}_4(y) = \emptyset$, we have $\mathcal{T}(y) = simple$, and

$y \in \mathcal{E}_3(x)$ from the meet definition; in the case that $\mathcal{E}_4(y) \neq \emptyset$, from the hypothesis we have $\mathcal{E}_3(y) \neq \emptyset$ and again $y \in \mathcal{E}_3(x)$ from the meet definition. Therefore, $y \in \mathcal{E}_3(x)$ as required.

It is trivial to see that $\mathbf{ST}_4 \leq \mathbf{ST}_3$ if $\mathbf{ST}_4 = \emptyset$. If $\mathbf{ST}_4 \neq \emptyset$, notice that $(\forall x \in X_4)\mathcal{E}_4(x) \subseteq \mathcal{E}_3(x)$ implies $X_4 = \mathcal{E}_4^*(x_o) \subseteq \mathcal{E}_3^*(x_o) = X_3$. Then $\mathbf{ST}_4 \leq \mathbf{ST}_3$ from Lemma 2.7.

Remarks

1. Two examples shown in Figure 2.7 demonstrate how to compute the meet of two state trees. Let $\mathbf{ST}_3 = \mathbf{ST}_1 \wedge \mathbf{ST}_2$.

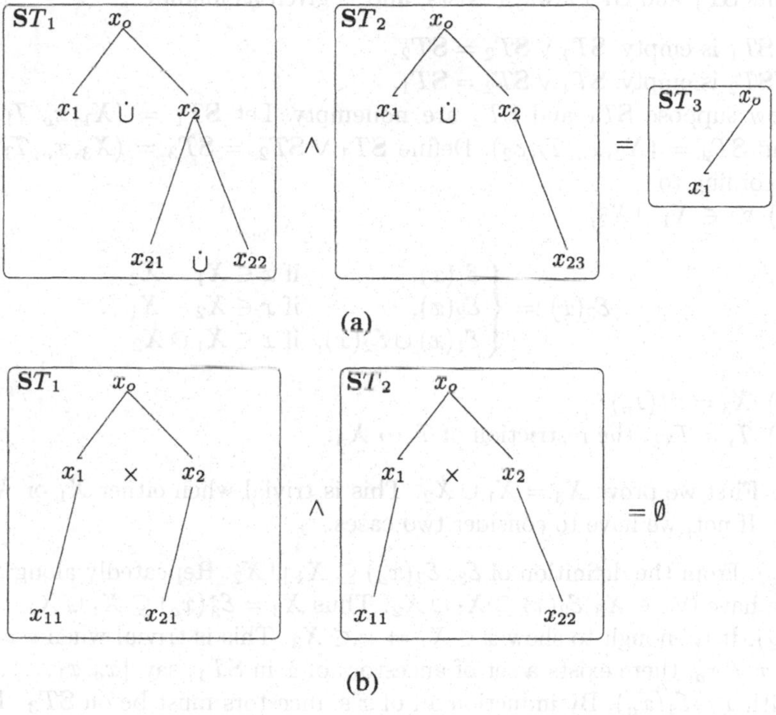

Fig. 2.7. Examples for the meet operator

a) In example (a), apply the definition of $\wedge$ for the case of OR superstates. We have $x_1 \in \mathcal{E}_3(x_o)$ because
   i. $x_1 \in \mathcal{E}_1(x_o) \cap \mathcal{E}_2(x_o)$, and
   ii. $\mathcal{T}(x_1) = \mathit{simple}$ and $\mathcal{E}_3(x_1) = \emptyset$ from the definition of $\wedge$ for the case of simple states.

   But $x_2 \notin \mathcal{E}_3(x_o)$ because even though

$$x_2 \in \mathcal{E}_1(x_o) \cap \mathcal{E}_2(x_o),$$

the statement ($\mathcal{T}(x_2) = or$ and $\mathcal{E}_3(x_2) = \emptyset$) violates condition (ii) since

$$\mathcal{E}_1(x_2) \cap \mathcal{E}_2(x_2) = \emptyset.$$

b) In example (b), apply the definition of $\wedge$ for the case of AND superstate. Because $\mathcal{E}_3(x_2) = \emptyset$, $\mathcal{E}_3(x_o) = \emptyset$ and therefore $\mathbf{ST}_1 \wedge \mathbf{ST}_2 = \emptyset$.

2. Notice that $X_3 \subseteq X_1 \cap X_2$. In the examples shown in Figure 2.7, $X_3 \subset X_1 \cap X_2$ (strict containment!). This is the reason that we cannot define $\wedge$ by simply saying $X_3 = X_1 \cap X_2$ where $\mathcal{E}_3, \mathcal{T}_3$ are the restriction functions of $\mathcal{E}, \mathcal{T}$ to X_3, respectively.

Proposition 2.2 [join] [9] In the poset $(\mathcal{ST}(\mathbf{ST}), \leq)$, the *join* $\mathbf{ST}_1 \vee \mathbf{ST}_2$ of elements $\mathbf{ST}_1$ and $\mathbf{ST}_2$ always exists, and is given as follows.

1. If $\mathbf{ST}_1$ is empty, $\mathbf{ST}_1 \vee \mathbf{ST}_2 = \mathbf{ST}_2$.
2. If $\mathbf{ST}_2$ is empty, $\mathbf{ST}_1 \vee \mathbf{ST}_2 = \mathbf{ST}_1$.
3. Now suppose $\mathbf{ST}_1$ and $\mathbf{ST}_2$ are nonempty. Let $\mathbf{ST}_1 = (X_1, x_o, \mathcal{T}_1, \mathcal{E}_1)$ and $\mathbf{ST}_2 = (X_2, x_o, \mathcal{T}_2, \mathcal{E}_2)$. Define $\mathbf{ST}_1 \vee \mathbf{ST}_2 = \mathbf{ST}_3 = (X_3, x_o, \mathcal{T}_3, \mathcal{E}_3)$ according to
   a) $\forall x \in X_1 \cup X_2$,
$$\mathcal{E}_3(x) := \begin{cases} \mathcal{E}_1(x), & \text{if } x \in X_1 - X_2 \\ \mathcal{E}_2(x), & \text{if } x \in X_2 - X_1 \\ \mathcal{E}_1(x) \cup \mathcal{E}_2(x), & \text{if } x \in X_1 \cap X_2 \end{cases}.$$
   b) $X_3 = \mathcal{E}_3^*(x_o)$.
   c) $\mathcal{T}_3 = \mathcal{T}_{X_3}$, the restriction of $\mathcal{T}$ to X_3.

Proof. First we prove $X_3 = X_1 \cup X_2$. This is trivial when either X_1 or X_2 is empty. If not, we have to consider two cases.

1. ($\subseteq$). From the definition of $\mathcal{E}_3$, $\mathcal{E}_3(x_o) \subseteq X_1 \cup X_2$. Repeatedly along ST_3, we have $(\forall x \in X_3)\mathcal{E}_3(x) \subseteq X_1 \cup X_2$. Thus $X_3 = \mathcal{E}_3^*(x_o) \subseteq X_1 \cup X_2$.
2. ($\supseteq$). It is enough to show $x \in X_1 \Rightarrow x \in X_3$. This is trivial when $x = x_o$. If $x \neq x_o$, there exists a set of ancestors of x in $\mathbf{ST}_1$, say $\{x_o, x_1, \ldots, x_n\}$ with $x \in \mathcal{E}_1(x_n)$. By induction all of x's ancestors must be on $\mathbf{ST}_3$. Thus $x \in X_3$ as required.

Now we prove $\mathbf{ST}_3 = \mathbf{ST}_1 \vee \mathbf{ST}_2$. Obviously $\mathbf{ST}_3$ is a sub-state-tree of $\mathbf{ST}$ because $X_3 = \mathcal{E}_3^*(x_o)$ and $(\forall x \in X_3)\mathcal{E}_3(x) \subseteq \mathcal{E}(x)$. We still need to show that

1. $\mathbf{ST}_1 \leq \mathbf{ST}_3$ and $\mathbf{ST}_2 \leq \mathbf{ST}_3$.
   This follows directly from Lemma 2.7 and the fact $X_3 = X_1 \cup X_2$.
2. If $\mathbf{ST}_4 \in \mathcal{ST}(\mathbf{ST})$ and $\mathbf{ST}_1 \leq \mathbf{ST}_4$, $\mathbf{ST}_2 \leq \mathbf{ST}_4$, then $\mathbf{ST}_3 \leq \mathbf{ST}_4$.
   This follows directly from Lemma 2.7 and the fact that $X_3 = X_1 \cup X_2 \subseteq X_4$.

[9] An example is given after the proposition.

Remarks

1. Figure 2.8 illustrates how to compute the join ($\vee$) of two state trees $\mathbf{ST}_1$ and $\mathbf{ST}_2$. It is simpler than $\wedge$ as we just need to place all states in $\mathbf{ST}_1$ and $\mathbf{ST}_2$ on the resulting state tree. The reason for the more complicated definition of $\wedge$ is that deleting a state from a sub-state-tree may result in a tree that is no longer a sub-state-tree of $\mathbf{ST}$ (too much may be cut), but adding a state will definitely guarantee that the tree is still a sub-state-tree of $\mathbf{ST}$.

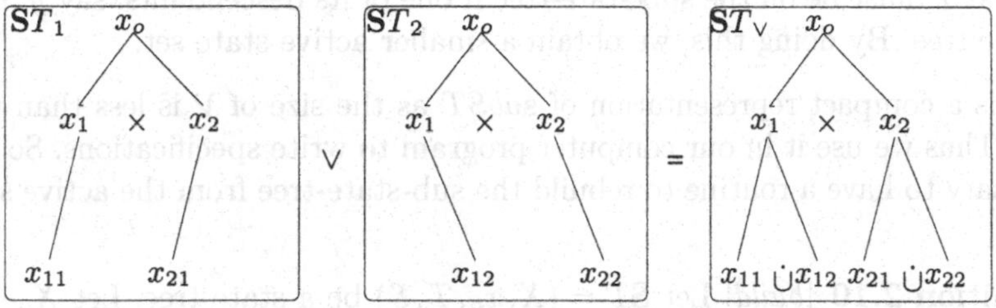

Fig. 2.8. $\mathbf{ST}_1 \vee \mathbf{ST}_2$

From Proposition 2.1 and 2.2, we have that $(\mathcal{ST}(\mathbf{ST}), \leq)$ is a *lattice*. In particular the bottom element is $\emptyset$ and the top element is $\mathbf{ST}$.

Given a state tree $\mathbf{ST}$, there are several ways to represent a sub-state-tree of $\mathbf{ST}$. The basic method is just to draw the whole picture. Another more efficient way is to use a *characteristic state set* to represent the sub-state-tree. Such states, called *active*, are defined as follows.

Definition 2.9 [active] Let $\mathbf{ST} = (X, x_o, T, \mathcal{E})$ be a state tree. Let $subST = (Y, x_o, T', \mathcal{E}')$ be a sub-state-tree of $\mathbf{ST}$. Let $z \in Y$ and $T'(z) = or$. We say $x \in \mathcal{E}'(z)$ is *active* if the following holds

$$\mathcal{E}'^*(x) = \mathcal{E}^*(x) \;\&\; \mathcal{E}'^*(z) \subset \mathcal{E}^*(z),$$

i.e., x is active if all of its descendants on $\mathbf{ST}$ are on the $subST$ but at least one descendant of z is not on the $subST$. Trivially, there is no active state for an empty sub-state-tree, and the only active state for $\mathbf{ST}$ itself is the root state x_o.

Denote by $\mathcal{V}(z)$ the set of all active states which expand z. Thus a proper sub-state-tree is uniquely represented by the *active state set* $\mathcal{V} = \bigcup_{\forall z \in Z} \mathcal{V}(z)$, where Z is the set of OR superstates that have active children.

◇

Remarks

1. For example, in Figure 2.6 on page 22, the active states of $\mathbf{ST}_1$ are x_2, x_{31}, x_{111}, and x_{121}. Intuitively, one can rebuild the whole sub-state-tree by adding all of their ancestors and descendants (if any). We will formalize the idea by introducing a *build* function later.

2. The active state set is a compact representation of a sub-state-tree. Based on the information stored in the active state set, we can rebuild the sub-state-tree. So we want the active state set to be as small as possible. This is the reason for choosing only OR components to be active, as we know (from the definition of sub-state-tree) that if an AND superstate is on a sub-state-tree then *all* of its children must be on the tree as well.
3. Assume $x < y$ and y is active. Then x is not active, from the definition (notice that $\mathcal{E}'^*(x) \subset \mathcal{E}^*(x)$ as y is active). That is, no ancestor of an active state can be active. The intuition behind this is that we can infer that x must be on the sub-state-tree if one of its descendants, say y, is on the tree. By doing this, we obtain a smaller active state set.

$\mathcal{V}$ is a compact representation of *subST* as the size of $\mathcal{V}$ is less than that of Y. Thus we use it in our computer program to write specifications. So it is necessary to have a routine to rebuild the sub-state-tree from the active state set $\mathcal{V}$.

Definition 2.10 [*build*] Let $\mathbf{ST} = (X, x_o, \mathcal{T}, \mathcal{E})$ be a state tree. Let $X_o \subseteq X$ be the set of all OR components. Then *build* $: \mathbf{ST} \times 2^{X_o} \longrightarrow \mathcal{ST}(\mathbf{ST})$ is a partial function that maps an active state set $\mathcal{V} \in 2^{X_o}$ to a sub-state-tree $\mathbf{ST}_\mathcal{V} = (X_\mathcal{V}, x_o, \mathcal{T}_\mathcal{V}, \mathcal{E}_\mathcal{V})$ of $\mathbf{ST}$, defined as follows.

1. $\mathit{build}(\mathbf{ST}, \emptyset) = \emptyset$, i.e., an empty active state set is mapped to an empty state tree.
2. $\mathit{build}(\mathbf{ST}, \{x_o\}) = \mathbf{ST}$, i.e., $\{x_o\}$ is mapped to $\mathbf{ST}$ itself.
3. Otherwise $x_o \in X_\mathcal{V}$ and we define $\mathcal{E}_\mathcal{V}$ recursively by
   a) $\mathcal{T}(x_o) = or$. Then $y \in \mathcal{E}_\mathcal{V}(x_o)$ iff
   $$\mathcal{E}^*(y) \cap \mathcal{V} \neq \emptyset.$$
   Recursively, for each $y \in \mathcal{E}_\mathcal{V}(x_o)$ and y is a superstate, define
   $$\mathit{build}(\mathbf{ST}^y, \mathcal{V} \cap \mathcal{E}^*(y))$$
   as the child state tree rooted by y that expands x_o in $\mathbf{ST}_\mathcal{V}$.
   b) $\mathcal{T}(x_o) = and$. Then $\mathcal{E}_\mathcal{V}(x_o) = \mathcal{E}(x_o)$ (required by the definition of sub-state-tree). For all $y \in \mathcal{E}_\mathcal{V}(x_o)$, define
   $$\begin{cases} \mathbf{ST}^y, & \text{if } \mathcal{V} \cap \mathcal{E}^*(y) = \emptyset \\ \mathit{build}(\mathbf{ST}^y, \mathcal{V} \cap \mathcal{E}^*(y)), & \text{if } \mathcal{V} \cap \mathcal{E}^*(y) \neq \emptyset \end{cases}$$
   as the child state tree rooted by y that expands x_o in $\mathbf{ST}_\mathcal{V}$.

   There are no other states in $X_\mathcal{V}$.

◇

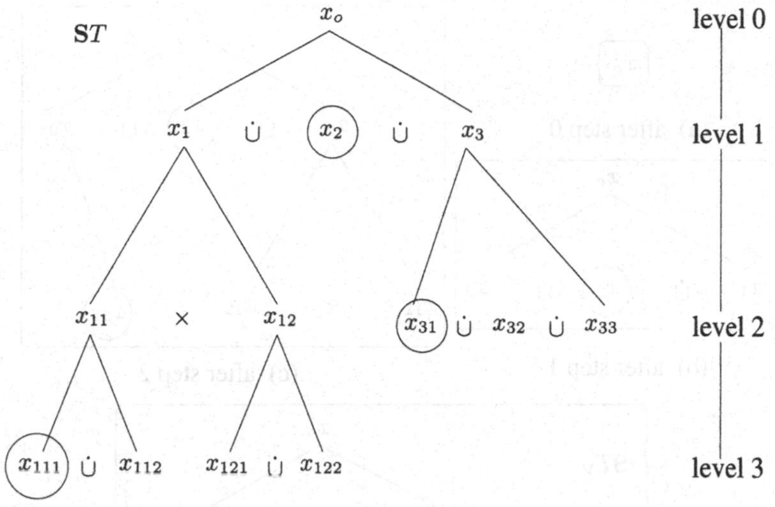

$$\mathcal{V} = \{x_{111}, x_2, x_{31}\}$$

Fig. 2.9. Example: *build*

Remarks

1. In Figure 2.9, there is a state tree $\mathbf{ST}$ and an active state set $\mathcal{V} = \{x_{111}, x_2, x_{31}\}$. We circle those active states on the $\mathbf{ST}$ so that the reader can easily find them. Now we demonstrate how the map *build* works in Figure 2.10.
   a) Step 0. As $\mathcal{V} \neq \emptyset$ and $\mathcal{V} \neq \{x_o\}$, we have $x_o \in X_\mathcal{V}$. See illustration in Figure 2.10 (a).
   b) Step 1. x_o is OR superstate. So we have
      i. $x_1 \in \mathcal{E}_\mathcal{V}(x_o)$ as $\mathcal{E}^*(x_1) \cap \mathcal{V} = \{x_{111}\} \neq \emptyset$;
      ii. $x_2 \in \mathcal{E}_\mathcal{V}(x_o)$ as $\mathcal{E}^*(x_2) \cap \mathcal{V} = \{x_2\} \neq \emptyset$;
      iii. $x_3 \in \mathcal{E}_\mathcal{V}(x_o)$ as $\mathcal{E}^*(x_3) \cap \mathcal{V} = \{x_{31}\} \neq \emptyset$.
   c) Step 2. x_1 and x_3 are superstates. So we call *build* on the child state trees rooted by x_1 and x_3, respectively. As x_1 is an AND superstate, we have $\mathcal{E}_\mathcal{V}(x_1) = \mathcal{E}(x_1)$. So x_{11} and x_{12} are on $\mathbf{ST}_\mathcal{V}$. As x_3 is an OR superstate, we apply the same operation on it as in step 1 to get only $x_{31} \in \mathcal{E}_\mathcal{V}(x_3)$.
   d) Step 3. Now check x_{11} and x_{12}. Notice that x_1 is an AND superstate. As $\mathcal{E}^*(x_{11}) \cap \mathcal{V} \neq \emptyset$, we call *build* on the child state tree $\mathbf{ST}^{x_{11}}$ and obtain $x_{111} \in \mathcal{E}_\mathcal{V}(x_{11})$. On the other hand, as $\mathcal{E}^*(x_{12}) \cap \mathcal{V} = \emptyset$, we know that all children of x_{12}, x_{121} and x_{122} are on the sub-state-tree $\mathbf{ST}_\mathcal{V}$.

   The resulting sub-state-tree $\mathbf{ST}_\mathcal{V}$ is uniquely determined by $\mathcal{V}$. We can tell that $\mathbf{ST}_\mathcal{V}$ will need much more computer storage than $\mathcal{V}$ as the size of $X_\mathcal{V}$ is much larger than that of $\mathcal{V}$.

The following function *count* is useful to measure the "size" of a state tree.

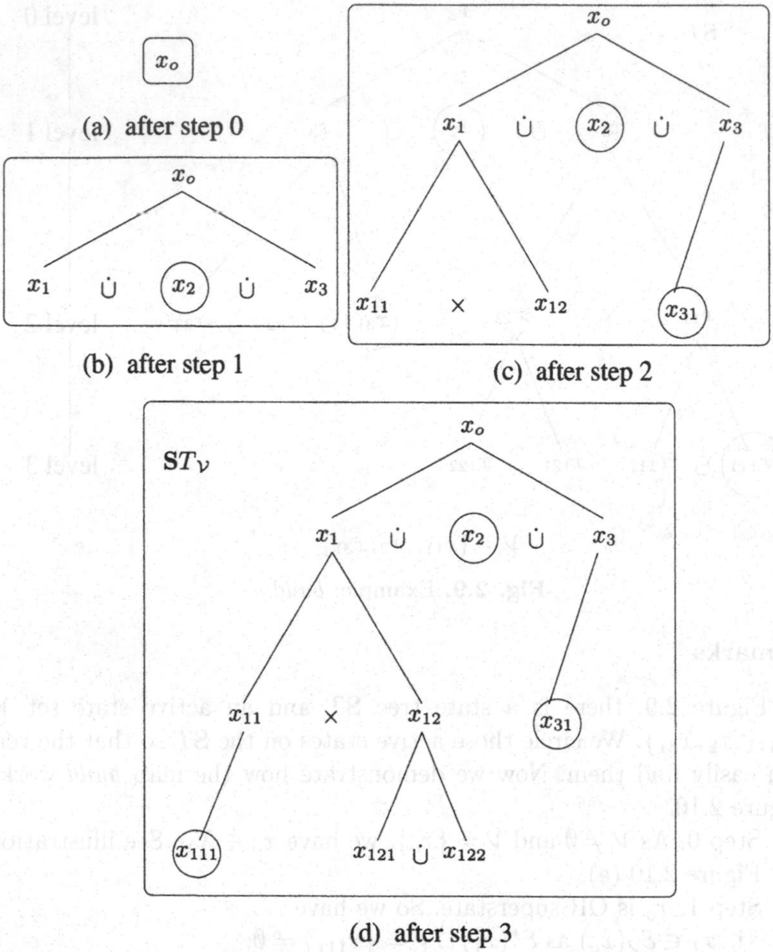

Fig. 2.10. How *build* works

Definition 2.11 [count] Let $\mathbf{ST} = (X, x_o, \mathcal{T}, \mathcal{E})$ be a state tree. For all $y \in \mathcal{E}(x_o)$ write $\mathbf{ST}^y = (\mathcal{E}^*(y), y, \mathcal{T}_{\mathcal{E}^*(y)}, \mathcal{E}_{\mathcal{E}^*(y)})$ for the child state trees of x_o. The *count* function of $\mathbf{ST}$ is defined recursively along the state tree $\mathbf{ST}$ according to

$$\text{count}(\mathbf{ST}) := \begin{cases} \prod_{\forall \mathbf{ST}^y} \text{count}(\mathbf{ST}^y), & \text{if } \mathcal{T}(x_o) = \textit{and} \\ \sum_{\forall \mathbf{ST}^y} \text{count}(\mathbf{ST}^y), & \text{if } \mathcal{T}(x_o) = \textit{or} \\ 1, & \text{if } \mathcal{T}(x_o) = \textit{simple} \end{cases}.$$

◊

In Figure 2.11, the count of $\mathbf{ST}$ is computed as follows.

$$\begin{aligned}\text{count}(\mathbf{ST}) &= \text{count}(\mathbf{ST}^{x_1}) + \text{count}(\mathbf{ST}^{x_2}) + \text{count}(\mathbf{ST}^{x_3}) \\ &= \text{count}(\mathbf{ST}^{x_{11}}) \times \text{count}(\mathbf{ST}^{x_{12}}) + 1 + 3 \\ &= 2 \times 2 + 4 \\ &= 8.\end{aligned}$$

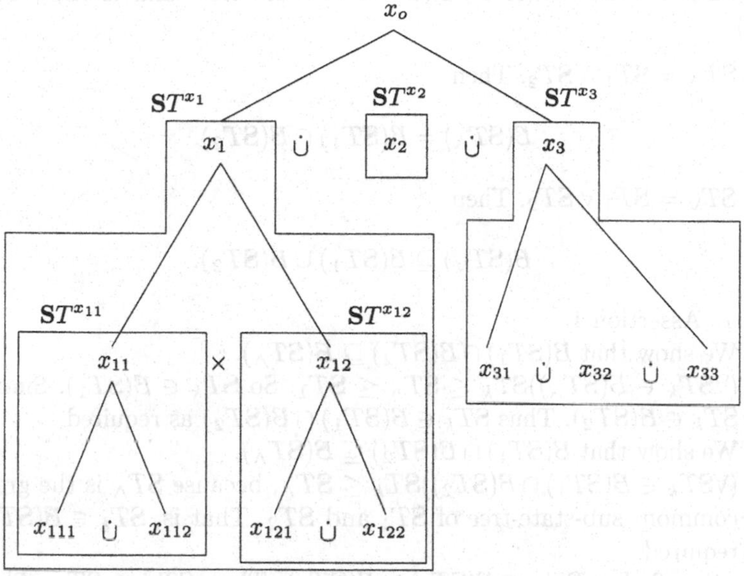

Fig. 2.11. count($\mathbf{ST}$) = 8

Definition 2.12 [basic sub-state-tree, basic state tree] Let $\mathbf{ST} = (X, x_o, T, \mathcal{E})$ be a state tree. Let $subST \in ST(\mathbf{ST})$. Say $subST$ is a *basic* sub-state-tree of $\mathbf{ST}$ if count($subST$) = 1; we call it a *basic state tree* if there is no ambiguity. Define

$$\mathcal{B}(\mathbf{ST}) := \{subST | \text{count}(subST) = 1\}$$

as the basic sub-state-tree set of $\mathbf{ST}$. Then $|\mathcal{B}(\mathbf{ST})| = \text{count}(\mathbf{ST})$, i.e., the number of basic sub-state-trees of $\mathbf{ST}$ is equal to count($\mathbf{ST}$).

◇

An important property of a basic state tree is that *each OR superstate on the tree must have exactly one child.* Otherwise its count will be greater than 1. In other words, no two states on a basic state tree can be exclusive.

In the original RW framework, the state space (of an automaton) is a set of elements labelled by the number set $\{0, 1, \ldots, n-1\}$, where n is the state space size. For convenience, we call the state in the RW framework a *flat state* or *RW state*. In our STS framework, the whole state space is "encoded" as a state tree, say $\mathbf{ST}$, with count($\mathbf{ST}$) $= n$. We could build the state space in the RW framework from $\mathbf{ST}$ by simply enumerating the elements in $\mathcal{B}(\mathbf{ST})$ and relabelling each element by a unique number. That is, an element in $\mathcal{B}(\mathbf{ST})$ is equivalent to a flat state in an automaton. So it will be useful to provide the relationship between operator $\wedge(\vee)$ on state trees and set operator $\cap(\cup)$ on the subsets of $\mathcal{B}(\mathbf{ST})$.

Lemma 2.9 Let $\mathbf{ST} = (X, x_o, \mathcal{T}, \mathcal{E})$ be a state tree, and let $\mathbf{ST}_1, \mathbf{ST}_2 \in \mathcal{ST}(\mathbf{ST})$.

1. Let $\mathbf{ST}_\wedge = \mathbf{ST}_1 \wedge \mathbf{ST}_2$. Then

$$\mathcal{B}(\mathbf{ST}_\wedge) = \mathcal{B}(\mathbf{ST}_1) \cap \mathcal{B}(\mathbf{ST}_2).$$

2. Let $\mathbf{ST}_\vee = \mathbf{ST}_1 \vee \mathbf{ST}_2$. Then

$$\mathcal{B}(\mathbf{ST}_\vee) \supseteq \mathcal{B}(\mathbf{ST}_1) \cup \mathcal{B}(\mathbf{ST}_2).$$

Proof. 1. Assertion 1.
   a) We show that $\mathcal{B}(\mathbf{ST}_1) \cap \mathcal{B}(\mathbf{ST}_2) \supseteq \mathcal{B}(\mathbf{ST}_\wedge)$.
   $(\forall \mathbf{ST}_k \in \mathcal{B}(\mathbf{ST}_\wedge))\mathbf{ST}_k \leq \mathbf{ST}_\wedge \leq \mathbf{ST}_1$. So $\mathbf{ST}_k \in \mathcal{B}(\mathbf{ST}_1)$. Similarly, $\mathbf{ST}_k \in \mathcal{B}(\mathbf{ST}_2)$. Thus $\mathbf{ST}_k \in \mathcal{B}(\mathbf{ST}_1) \cap \mathcal{B}(\mathbf{ST}_2)$ as required.
   b) We show that $\mathcal{B}(\mathbf{ST}_1) \cap \mathcal{B}(\mathbf{ST}_2) \subseteq \mathcal{B}(\mathbf{ST}_\wedge)$.
   $(\forall \mathbf{ST}_k \in \mathcal{B}(\mathbf{ST}_1) \cap \mathcal{B}(\mathbf{ST}_2))\mathbf{ST}_k \leq \mathbf{ST}_\wedge$, because $\mathbf{ST}_\wedge$ is the greatest common sub-state-tree of $\mathbf{ST}_1$ and $\mathbf{ST}_2$. That is, $\mathbf{ST}_k \in \mathcal{B}(\mathbf{ST}_\wedge)$ as required.
2. Assertion 2. Let $\mathbf{ST}_k \in \mathcal{B}(\mathbf{ST}_1) \cup \mathcal{B}(\mathbf{ST}_2)$. Then $\mathbf{ST}_k \leq \mathbf{ST}_\vee$. That is, $\mathbf{ST}_k \in \mathcal{B}(\mathbf{ST}_\vee)$ as required.

Remarks

1. The equality for the operator $\wedge$ provides great power in computation, as usually a state tree $\mathbf{ST}$ is much simpler than its basic sub-state-tree set $\mathcal{B}(\mathbf{ST})$, which can contain billions of basic state trees in a moderately complex system. We say $\mathbf{ST}$ is a *symbol* for the set $\mathcal{B}(\mathbf{ST})$. Notice that each element in $\mathcal{B}(\mathbf{ST})$ is equivalent to a flat state in the RW framework. Thus, the intersection operations over special sets of the form $\mathcal{B}(\mathbf{ST}_1)$ and $\mathcal{B}(\mathbf{ST}_2)$ can be done *symbolically* by $\mathbf{ST}_1 \wedge \mathbf{ST}_2$ and (the best part) the result is represented economically by $\mathbf{ST}_\wedge$.
2. Figure 2.12 is a counterexample to explain why we cannot have $=$ instead of $\subseteq$ for the operator $\vee$. Notice that both $\mathbf{ST}_1$ and $\mathbf{ST}_2$ are basic sub-state-trees and $\mathbf{ST}_\vee$ has 4 basic sub-state-trees. Thus $\mathcal{B}(\mathbf{ST}_1) \cup \mathcal{B}(\mathbf{ST}_2) \subset \mathcal{B}(\mathbf{ST}_\vee)$.
3. Only under very strong conditions can we change the inequality to equality for the operator $\vee$. That means that we have to use more expensive ways to compute and represent the union of two sets of RW states. Unfortunately,

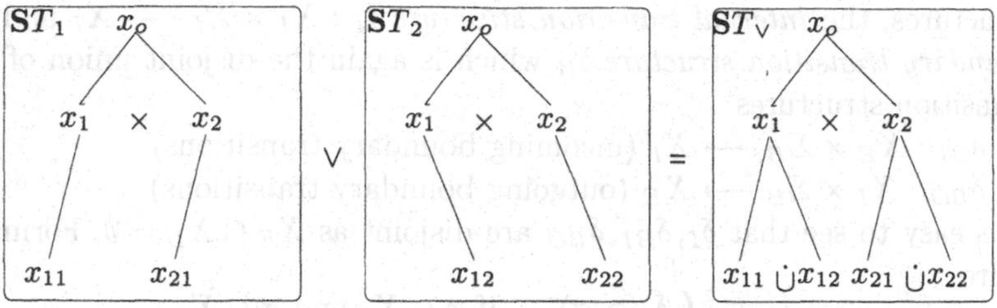

Fig. 2.12. $\mathcal{B}(\mathbf{S}T_1) \cup \mathcal{B}(\mathbf{S}T_2) \subset \mathcal{B}(\mathbf{S}T_\vee)$

set union is the fundamental operation in the fixpoint computation of our optimal nonblocking supervisor. This can make the synthesis of STS computationally very intensive. We attack this problem by introducing BDD, explained in later chapters.

In this section, the concepts of state tree and sub-state-tree were introduced to describe the hierarchical state space. We proved that $(ST(ST), \leq)$ is a lattice, namely the meet and join of any two sub-state-trees always exist. This is a fundamental result because it underlies the definition of our global transition function Δ in section 2.3.

2.2 Holon

If we assign transition structure to $\mathcal{B}(\mathbf{S}T)$, taking each basic sub-state-tree as a flat state of RW framework, we can define the *global* behavior of the system "flatly". However, the set $\mathcal{B}(\mathbf{S}T)$ is so large for complex systems that the assignment may be infeasible to carry out. Instead, we introduce the concept of *holon*, the *local* behavior, and then build the global behavior structurally.

Definition 2.13 [Holon] A *Holon* H is defined as a 5-tuple

$$H := (X, \Sigma, \delta, X_o, X_m),$$

where

- X is the nonempty state set, structured as the disjoint union of the (possibly empty) *external state set* X_E and the nonempty *internal state set* X_I, i.e., $X = X_E \dot\cup X_I$ with $X_E \cap X_I = \emptyset$.
- Σ is the event set, structured as the disjoint union of the *boundary event set* Σ_B and the *internal event set* Σ_I, i.e., $\Sigma = \Sigma_B \cup \Sigma_I$ with $\Sigma_B \cap \Sigma_I = \emptyset$. Σ can also be partitioned by the sets of controllable and uncontrollable events, i.e., $\Sigma = \Sigma_c \dot\cup \Sigma_u$ with $\Sigma_c \cap \Sigma_u = \emptyset$.
- The transition structure $\delta : X \times \Sigma \longrightarrow X$ is a partial function. Write $\delta(x, \sigma)!$ if $\delta(x, \sigma)$ is defined. δ is the disjoint union [10] of two transition

[10] We say two transition structures $\delta_i : X \times \Sigma \longrightarrow X, i = 1, 2$ are disjoint if the tuple sets $\{(x, \sigma, \delta_1(x, \sigma)) | \delta_1(x, \sigma)!\}$ and $\{(x, \sigma, \delta_2(x, \sigma)) | \delta_2(x, \sigma)!\}$ are disjoint.

structures, the *internal transition structure* $\delta_I : X_I \times \Sigma_I \longrightarrow X_I$ and the *boundary transition structure* δ_B which is again the disjoint union of two transition structures

– $\delta_{BI} : X_E \times \Sigma_B \longrightarrow X_I$ (incoming boundary transitions)
– $\delta_{BO} : X_I \times \Sigma_B \longrightarrow X_E$ (outgoing boundary transitions)

It is easy to see that $\delta_I, \delta_{BI}, \delta_{BO}$ are disjoint as $X_E \cap X_I = \emptyset$. Formally, write

$$\delta(x,\sigma) := \begin{cases} \delta_I(x,\sigma), & \text{if } x \in X_I \text{ and } \sigma \in \Sigma_I \\ \delta_{BI}(x,\sigma), & \text{if } x \in X_E \text{ and } \sigma \in \Sigma_B \\ \delta_{BO}(x,\sigma), & \text{if } x \in X_I \text{ and } \sigma \in \Sigma_B \end{cases}.$$

Also we require that the transition structure be deterministic with respect to transition event labelling, i.e.,

$$(\forall x, y, y' \in X)(\forall \sigma \in \Sigma)(\delta(x,\sigma) = y \,\&\, \delta(x,\sigma) = y') \Rightarrow y = y'.$$

- $X_o \subseteq X_I$ is the *initial state set* [11], where X_o has exactly the target states of incoming boundary transitions if δ_{BI} is defined. Otherwise X_o is a nonempty subset of X_I selected according to convenience. [12] Formally, let
$$U := \{\delta_{BI}(x,\sigma) | (\exists x \in X_E, \sigma \in \Sigma_B)\delta_{BI}(x,\sigma)!\}.$$
We have
$$X_o := \begin{cases} U, & \text{if } U \neq \emptyset \\ Z, \text{ where } \emptyset \subset Z \subseteq X_I, & \text{if } U = \emptyset \end{cases}.$$
From now on we write $\delta_{BI} : X_E \times \Sigma_B \longrightarrow X_o$ (pfn) .
- $X_m \subseteq X_I$ is the *terminal state set*, where X_m has exactly the source states of the outgoing boundary transitions if δ_{BO} is defined. Otherwise X_m is a selected nonempty subset of X_I. Formally, let
$$U := \{x | x \in X_I \,\&\, (\exists \sigma \in \Sigma_B)\delta_{BO}(x,\sigma)!\}.$$
We have
$$X_m := \begin{cases} U, & \text{if } U \neq \emptyset \\ Z, \text{ where } \emptyset \subset Z \subseteq X_I, & \text{if } U = \emptyset \end{cases}.$$
Similarly $\delta_{BO} : X_m \times \Sigma_B \longrightarrow X_E$ (pfn).

◊

Remarks

1. A holon $H := (X, \Sigma, \delta, X_o, X_m)$, shown in Figure 2.13, provides the general idea of what a holon looks like. In Figure 2.13, we have

[11] Notice that in Wang's thesis [Wan95, page 31], $X_o \subseteq X_E$ if X_E is nonempty, whereas X_o has exactly one element of X_I if X_E is empty. That is a bit confusing because X_o could be in X_E or X_I. Here we require it must be in X_I.

[12] Notice that this is more general than the initial state of a generator in original RW framework. We allow more than one initial state in our framework.

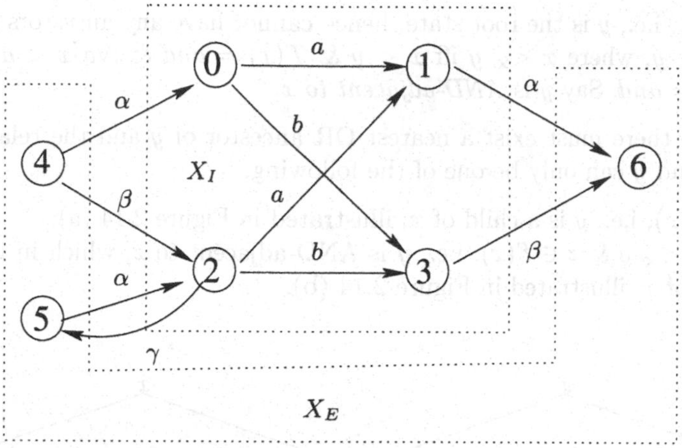

Fig. 2.13. A Holon H

- $X_E = \{4,5,6\}$ and $X_I = \{0,1,2,3\}$;
- $\Sigma_B = \{\alpha, \beta, \gamma\}$ and $\Sigma_I = \{a, b\}$;
- $X_o = \{0,2\}$;
- $X_m = \{1,2,3\}$.

Notice that $X_o \cap X_m = \{2\} \neq \emptyset$.

2. In a holon, the state set, event set, and transition structure are all partitioned. All of them give a holon two disjoint parts, an interface featured by boundary transition structure and an inner system featured by internal transition structure. Usually, a good model will have the inner system more complex than the interface.
3. Notice that, in the original RW framework, an automaton can have only one initial state. Here a holon is allowed to have more than one initial state, a feature that provides modelling flexibility to be exploited later (Chapter 3).
4. From its behavior, it is seen that a holon can be taken as a generalized generator (automaton) with more than one initial state.

2.3 State Tree Structure

Having *state tree* to represent the state space and *holons* to represent the local dynamics, now it is time to bring them together and build our STS model. First of all, we connect these two concepts by *matching* a holon, the local dynamics, to an *OR* superstate on the state tree.

Definition 2.14 [match] Let $\mathbf{ST} = (X, x_o, \mathcal{T}, \mathcal{E})$ be a state tree. Let $y \in X$ and $\mathcal{T}(y) = or$. Let $x \in X$ be the *nearest OR ancestor* of y , i.e., $x < y \;\&\; \mathcal{T}(x) = or$ and $((\forall z \in X) z < y \;\&\; \mathcal{T}(z) = or \Rightarrow z \leq x)$. Such x does not exist in the following two cases.

1. $y = x_o$, i.e., y is the root state, hence cannot have any ancestors at all.
2. $x_o <_\times y$, where $x <_\times y$ iff $x < y$ & $\mathcal{T}(x) = and$ & $(\forall a) x < a < y \Rightarrow \mathcal{T}(a) = and$. Say y is *AND-adjacent to* x.

Otherwise there must exist a nearest OR ancestor of y and the relation between x and y can only be one of the following,

1. $y \in \mathcal{E}(x)$, i.e., y is a child of x, illustrated in Figure 2.14 (a);
2. $(\exists z) z <_\times y$ & $z \in \mathcal{E}(x)$, i.e., y is AND-adjacent to z, which in turn is a child of x, illustrated in Figure 2.14 (b).

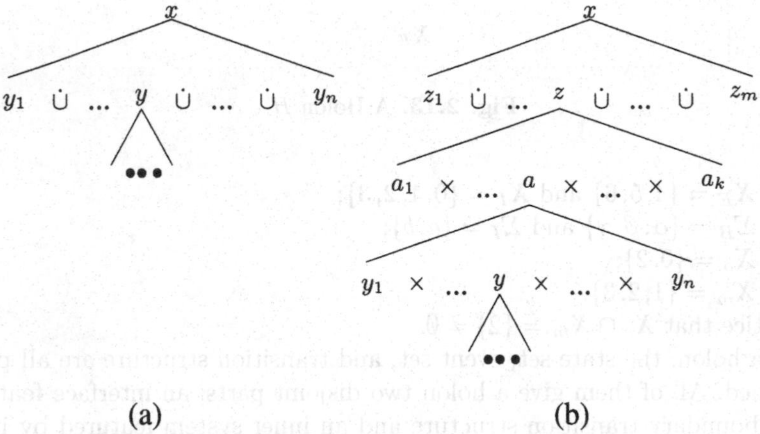

Fig. 2.14. Relation between y and its nearest OR ancestor x

A holon $H^y = (X_E^y \cup X_I^y, \Sigma^y, \delta^y, X_o^y, X_m^y)$ is said to be *matched* to the superstate y if

- internal structure matches.

$$X_I^y = \mathcal{E}(y)$$
$$X_o^y \subseteq \mathcal{E}(y)$$
$$X_m^y \subseteq \mathcal{E}(y)$$
$$\delta_I^y \ : \ X_I^y \times \Sigma_I^y \longrightarrow X_I^y$$

- external structure matches. (Namely x is the nearest OR ancestor of y; and $z <_\times y$ in case (b) of Figure 2.14.)

$$\begin{cases} X_E^y = \emptyset, & \text{if } x \text{ does not exist} \\ X_E^y \subseteq \mathcal{E}(x) - \{y\}, & \text{if } x \text{ exists and } y \in \mathcal{E}(x) \text{ as in (a) of Figure 2.14} \\ X_E^y \subseteq \mathcal{E}(x) - \{z\}, & \text{if } x \text{ exists and } z <_\times y \text{ as in (b) of Figure 2.14} \end{cases}$$
$$\delta_{BI} : X_E^y \times \Sigma_B^y \longrightarrow X_o^y$$
$$\delta_{BO} : X_m^y \times \Sigma_B^y \longrightarrow X_E^y$$

Suppose a holon H^x is matched to the OR superstate x. We say H^x is the *parent holon* of H^y and H^y is the *child holon* of H^x.

◊

Remarks

1. From the external structure match, if x does exist, $X_E^y \subset \mathcal{E}(x)$ implies that the superstate y cannot be a boundary state in $(X_o^x \cup X_m^x)$, i.e., all boundary states of matched holons must be simple states. This feature restricts direct vertical communication to parent/child holon pairs.
   Figure 2.15 illustrates the above statement. We borrow the graphical notation of statecharts to draw the state tree and its matched holons in (b). Notice that $X_E^y = \{p, q\} \not\subset \mathcal{E}(x) = \{q, y\}$ in (b). So the holon in (b) does not *match* y correctly. The reason is that the superstate y is an initial state of the holon H^x matched to x. To correct the problem, we need only add a simple state t between p and s to have the correct *match* of holons in (d), because now $X_E^y = \{q, t\} \subset \mathcal{E}(x) = \{q, y, t\}$.

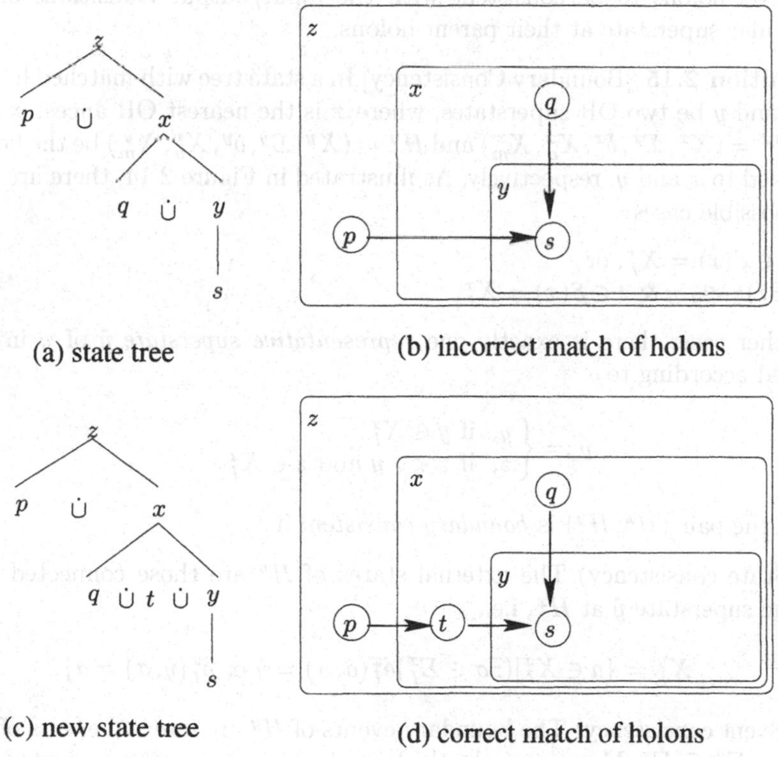
z
p
x
q
y
s
(a) state tree
z
x
q
y
p
s
(b) incorrect match of holons
z
p
x
q
t
y
s
(c) new state tree
z
x
q
y
p
t
s
(d) correct match of holons

Fig. 2.15. Example: no boundary states can be superstates

Notice that this is the only way allowed in our framework to assign transitions, i.e., no simple state or AND superstate can be matched by a holon. This fact is explained as follows:

1. A simple state, say y, has no nonempty state set to further expand it, i.e., $\mathcal{E}(y) = \emptyset$. So the only available holon that *could match* y would need an *empty* internal state set (from the above definition of internal structure matches), which is impossible because every holon must have *nonempty* internal state set.
2. The semantics of an AND superstate y is the AND of all children of y, namely to be inside y the system must be at *all* children of y *simultaneously*. However, any two states in a holon must be *exclusive*, i.e., in each run the holon must visit exactly one state at any particular moment. So it does not make sense to assign transitions in a holon to an AND superstate y, in which any two children of y are *parallel*.

We assign dynamics to a state tree by matching holons to all of the OR superstates on the tree. Of course, we need the boundary transitions of any low level holons to be consistent with the input/output transitions of the particular superstate at their parent holons.

Definition 2.15 [Boundary Consistency] In a state tree with matched holons, let x and y be two OR superstates, where x is the nearest OR ancestor of y. Let $H^x = (X^x, \Sigma^x, \delta^x, X_o^x, X_m^x)$ and $H^y = (X^y, \Sigma^y, \delta^y, X_o^y, X_m^y)$ be the holons matched to x and y, respectively. As illustrated in Figure 2.14, there are only two possible cases

1. $y \in \mathcal{E}(x) = X_I^x$, or
2. $(\exists z) z <_\times y \;\&\; z \in \mathcal{E}(x) = X_I^x$.

In either case, there is exactly one *representative superstate* $\hat{y}$ of y in X_I^x, defined according to

$$\hat{y} := \begin{cases} y, & \text{if } y \in X_I^x \\ z, & \text{if } z <_\times y \text{ and } z \in X_I^x \end{cases}.$$

Then the pair (H^x, H^y) is *boundary consistent* if

- (State consistency) The external states of H^y are those connected with the superstate $\hat{y}$ at H^x, i.e.,

$$X_E^y = \{a \in X_I^x | (\exists \sigma \in \Sigma_I^x) \delta_I^x(a, \sigma) = \hat{y} \text{ or } \delta_I^x(\hat{y}, \sigma) = a\}.$$

- (Event consistency) The boundary events of H^y are internal events of H^x, i.e., $\Sigma_B^y \subseteq \Sigma_I^x$. More precisely, the boundary events are those events which point to or exit from the superstate $\hat{y}$ at H^x, i.e.,

$$\Sigma_B^y = \{\sigma \in \Sigma_I^x | (\exists a \in X_I^x) \delta_I^x(a, \sigma) = \hat{y} \text{ or } \delta_I^x(\hat{y}, \sigma) = a\}.$$

- (Incoming boundary transition consistency) The incoming boundary transitions of H^y are consistent with those of the superstate $\hat{y}$ at H^x, i.e.,

$$(\forall a \in X_E^y, \sigma \in \Sigma_B^y)(\exists b \in X_o^y)(\delta_{BI}^y(a,\sigma) = b \text{ iff } \delta_I^x(a,\sigma) = \hat{y}).$$

- (Outgoing boundary transition consistency) The outgoing boundary transitions of H^y are consistent with those of the superstate $\hat{y}$ at H^x, i.e.,

$$(\forall b \in X_E^y, \sigma \in \Sigma_B^y)(\exists a \in X_m^y)(\delta_{BO}^y(a,\sigma) = b \text{ iff } \delta_I^x(\hat{y},\sigma) = b).$$

We also say holon H^y *expands* holon H^x if the pair is boundary consistent.

◊

Remarks

1. Figure 2.16 illustrates the boundary consistency between two holons. Let H^x be the holon matched to the superstate x. Let H_1^y and H_2^y be the two holons matched to the child state y of x. Notice that H_1^y expands H^x because the pair is boundary consistent. However H_2^y does not expand H^x because the pair does not satisfy the incoming boundary transition consistency, i.e., the transition $\delta_I^x(a,\alpha) = y$ of the high level holon H^x is not consistent with the transition $\delta_{BI}^y(a,\gamma) = 0$ of the low level holon H_2^y.
2. Even though the consistency may seem complicated, it is easy to verify. Intuitively, boundary consistency means that one can "plug" the low level holon into the high level holon *without* changing the input/output transitions of the former's representative superstate in the high level. This is illustrated in Figure 2.17.

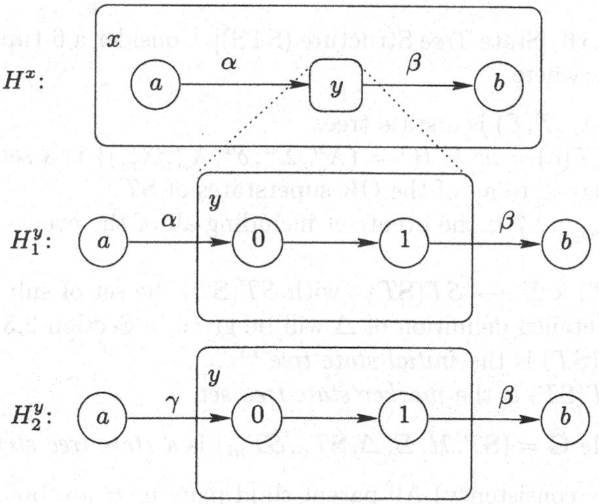

Fig. 2.16. Boundary consistency

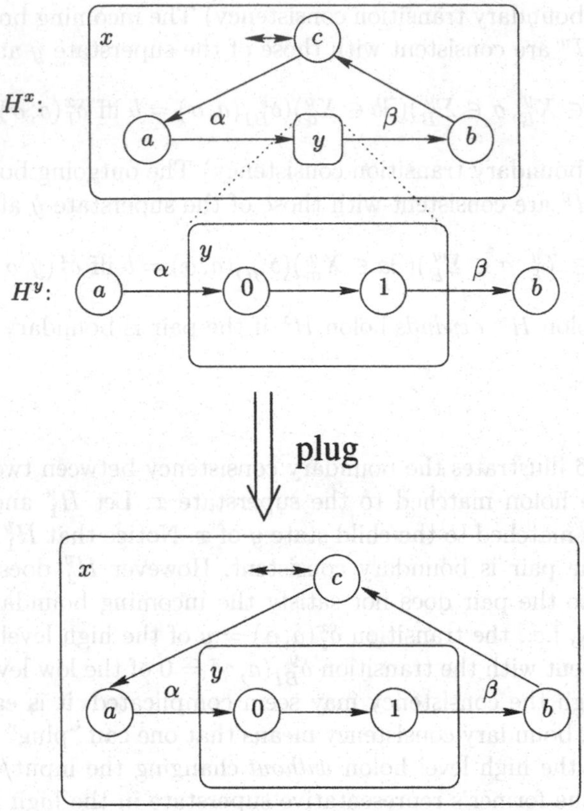

Fig. 2.17. "Plug"

Definition 2.16 [State Tree Structure (STS)] Consider a 6-tuple $(\mathbf{ST}, \mathcal{H}, \Sigma, \Delta, \mathbf{ST}_o, \mathcal{ST}_m)$, where

- $\mathbf{ST} := (X, x_o, \mathcal{T}, \mathcal{E})$ is a state tree.
- $\mathcal{H} := \{H^a | \mathcal{T}(a) = or \ \& \ H^a = (X^a, \Sigma^a, \delta^a, X^a_o, X^a_m)\}$ is a set of matching holons assigned to all of the OR superstates of $\mathbf{ST}$.
- $\Sigma := \bigcup_{\forall H^a \in \mathcal{H}} \Sigma^a_I$ is the event set including all of the events appearing in $\mathcal{H}$.
- $\Delta : \mathcal{ST}(\mathbf{ST}) \times \Sigma \longrightarrow \mathcal{ST}(\mathbf{ST})$, with $\mathcal{ST}(\mathbf{ST})$ the set of sub-state-trees of $\mathbf{ST}$. The detailed definition of Δ will be given in Section 2.3.1 below.
- $\mathbf{ST}_o \in \mathcal{ST}(\mathbf{ST})$ is the *initial state tree* [13].
- $\mathcal{ST}_m \subseteq \mathcal{ST}(\mathbf{ST})$ is the *marker state tree set.*

The 6-tuple $\mathbf{G} = (\mathbf{ST}, \mathcal{H}, \Sigma, \Delta, \mathbf{ST}_o, \mathcal{ST}_m)$ is a *state tree structure* if

1. (Boundary consistency) All parent-child pairs in $\mathcal{H}$ are boundary consistent, and

[13] Notice that its count may be greater than 1.

2. (Local coupling) Events of an inner transition structure can only be shared among those holons matched to the OR states that are AND-adjacent to the same AND superstate. Formally, for all superstates $a \neq b$ with matching holons $H^a, H^b \in \mathcal{H}$, we require

$$\Sigma_I^a \cap \Sigma_I^b \neq \emptyset \Rightarrow (\exists z) z <_\times a \;\&\; z <_\times b.$$

G is *well-formed* if **ST** is a well-formed state tree and **G** satisfies both boundary consistency and local coupling. **G** is *deterministic* if all of its holons are deterministic.

$\mathbf{G}^y = (\mathbf{ST}^y, \mathcal{H}^y, \Sigma^{\mathcal{H}^y}, \Delta^y, \mathbf{ST}^y_o, \mathcal{ST}^y_m)$ is a *child state tree structure* of x_o in **G** if it is also boundary consistent and locally coupled, where

- $\mathbf{ST}^y$ is a child state tree of x_o in **ST**, rooted by a superstate $y \in \mathcal{E}(x_o)$;
- $\mathcal{H}^y \subseteq \mathcal{H}$ is the set of holons matched to all of the OR superstates in $\mathbf{ST}^y$;
- $\Sigma^{\mathcal{H}^y}$ is the set of events in $\mathcal{H}^y$;
- $\Delta^y : \mathcal{ST}(\mathbf{ST}^y) \times \Sigma^{\mathcal{H}^y} \longrightarrow \mathcal{ST}(\mathbf{ST}^y)$, with $\mathcal{ST}(\mathbf{ST}^y)$ the set of sub-state-trees of $\mathbf{ST}^y$;
- $\mathbf{ST}^y_o \in \mathcal{ST}(\mathbf{ST}^y)$ is a child state tree of x_o in $\mathbf{ST}_o$, rooted by y;
- $\mathcal{ST}^y_m \subseteq \mathcal{ST}(\mathbf{ST}^y)$ has each of its elements a child state tree of x_o in its corresponding element of $\mathcal{ST}_m$, rooted by y.

We say that the set of STS $\{\mathbf{G}^y | y \in \mathcal{E}(x_o) \;\&\; \mathbf{G}^y$ is a child STS of x_o in $\mathbf{G}\}$ *expands* x_o.

◊

Remarks

1. One of the most appealing features of our state tree structures is its graphical representation (adapted from [Har87]). We can draw a system with billions of RW states on a few pages. Table 2.1 lists the new graphical notation for a system modelled in STS. Notice that we still keep using the same notation as in a flat model.
   The reason that we include item 2 is that sometimes the number of states is too large to draw all of them on one page. In that case, we can use the notation in item 2 to draw a superstate on the summary page and then draw its inner transition structure on another page. Also notice that we can always omit state labels to save space if there is no ambiguity.
2. A familiar example is the Small Factory with a buffer (from chapter 3 of [Won04]). Without hierarchy, the STS model of this particular example is almost the same as the synchronous product model in the original RW framework, with the exception of the new graphical notation for superstates. This confirms that our STS model is a natural extension of the RW framework. The superstates in this example are the AND superstate *Small Factory* and three OR superstates *M1*, *M2* and *BUF*.
3. Based on the STS model in Figure 2.18, we can write a specification and then design its controller. Suppose we are not satisfied with the result

Table 2.1. New graphical notation of STS

Item	Graphical Notation	Explanation
1		x is a simple state.
2	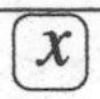	x is a superstate.
3	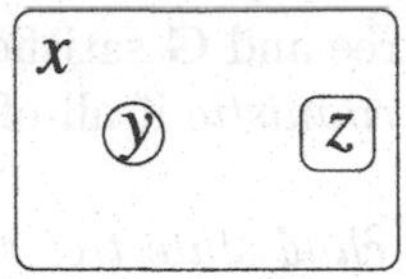	x is an OR superstate. It has two children, simple state y and another superstate z.
4	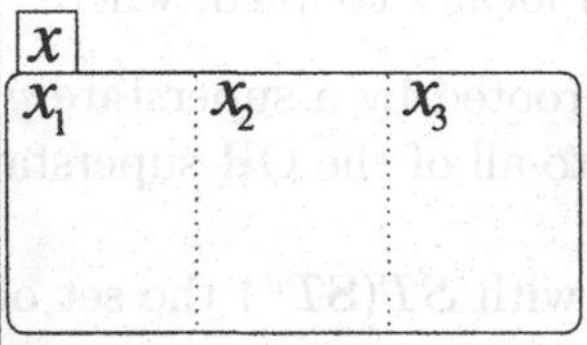	x is an AND superstate. It has three children, x_1, x_2 and x_3. All of them are required to be OR superstates

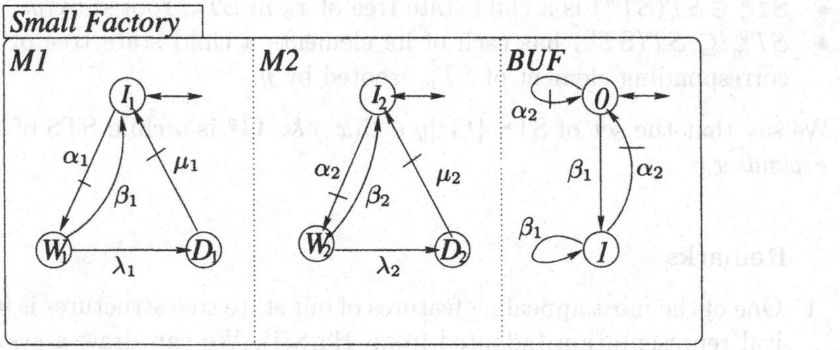

Fig. 2.18. Small Factory

because the controller is too restrictive. One solution is to look into the details of the system and add more information to the STS model. Suppose some controllable events can happen *inside* the state $W_i, i = 1, 2$, i.e., W_1 and W_2 are superstates now. We will have the Small Factory II, shown in Figure 2.19. The synthesis will then generate a less restrictive controller as it is possible now to disable events *inside* the state W_1 and W_2. Sometimes it is difficult to understand a complex system if we draw all of its holons on one page. An alternative is to draw the top level holons first, with arc rectangular boxes representing superstates, then draw the holons matched to those superstates one by one recursively. This way we can enjoy a top-down view of the whole system. Figure 2.20 is an alternative to the model drawn in Figure 2.19, where (a) shows the top level holons, (b) the holon matched to W_1 and (c) the holon matched to W_2. Later on we will use both methods to draw STS models. The advantage of the first

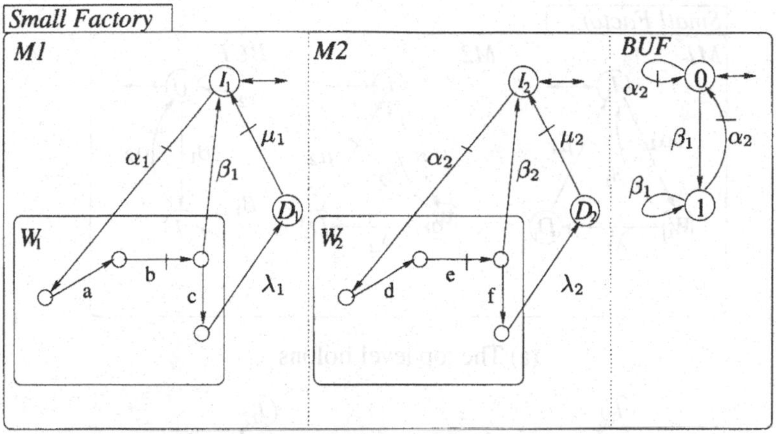

Fig. 2.19. Small Factory II

one is that the whole system is on one page and we can quickly find the information we want, whereas that of the second one is that we can easily add new details without redrawing the whole picture of the system. The authors prefer the second one because it can handle larger hierarchical systems.

4. In the Small Factory, suppose there are two independent jobs to be done before each broken machine is repaired. We can add this information into our STS model by modelling the states $D_i, i = 1, 2$ as AND superstates, as shown in Figure 2.21. Two holons, matching the superstates J_{11} and J_{12}, are drawn in one picture (d) to save space, because both share the boundary information of D_1. The way λ_1 and μ_1 are drawn in (d) reflects the fact that the system has to enter and leave J_{11} and J_{12} concurrently.

The function Δ is important, as it describes the *global* behavior of STS. In the next subsection, we are going to define the function Δ.

2.3.1 $\Delta : ST(ST) \times \Sigma \longrightarrow ST(ST)$

Let us first describe the semantics of a transition in STS. Basically we need to answer the following question:

What is the meaning of a transition with superstates as its source state or target state?

As we saw before, a transition connecting two simple states means that whenever the event occurs, the system will move instantly to the target state from the source state. As STS is a natural extension of the automaton model, we apply the same semantics to the STS too. In addition, we need to cover new cases where at least one state involved in a transition is a superstate. We find that one δ function is not enough to completely describe the meaning of

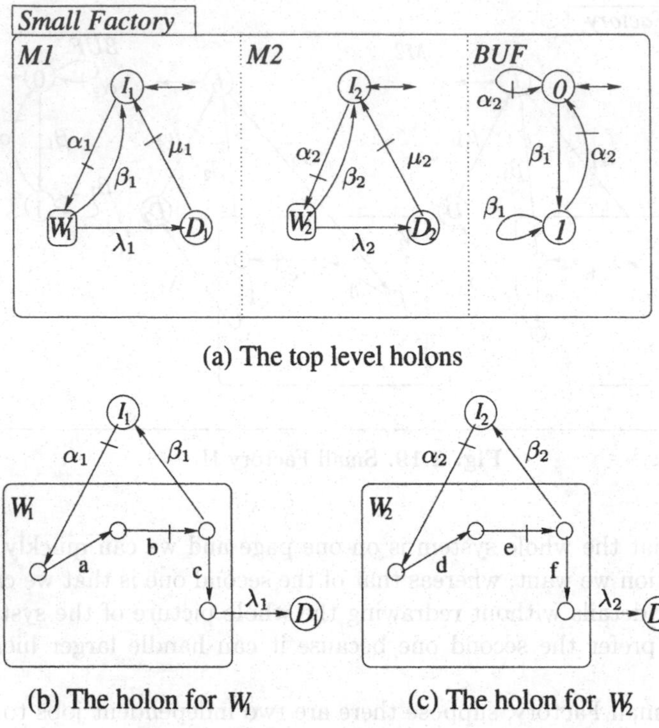

Fig. 2.20. Small Factory II: Alternative

a transition. For example, in Figure 2.22, a transition labelled with event α connects two simple states p_o and q_o. There are three δ functions associated with this transition:

1. $\delta_I^x(p, \alpha) = q$ (from the internal transition structure of holon H^x),
2. $\delta_{BO}^p(p_o, \alpha) = q$ (from the outgoing boundary transition structure of holon H^p), and
3. $\delta_{BI}^q(p, \alpha) = q_o$ (from the incoming boundary transition structure of holon H^q).

None of them by itself is adequate to describe the transition that moves the system from p_o to q_o. Precisely, the transition α transforms the (child) sub-state-tree $\mathbf{ST}_p^x$ [14] to another (child) sub-state-tree $\mathbf{ST}_q^x$, shown in Figure 2.23.

So what is needed is an extended function $\bar{\delta_I}$ of δ_I that specifies how to transform a (child) sub-state-tree into another one.

Definition 2.17 $[\bar{\delta_I}]$ Let $H^x = (X^x, \Sigma^x, \delta^x, X_o^x, X_m^x)$ be the holon matched to the OR superstate x in our STS. $\bar{\delta_I^x} : \mathcal{B}(\mathbf{ST}^x) \times \Sigma_I^x \longrightarrow \mathcal{B}(\mathbf{ST}^x)$ is a map described as follows.

[14] Recall in our notation, ST^x means the child state tree of $\mathbf{ST}$ with the root state x.

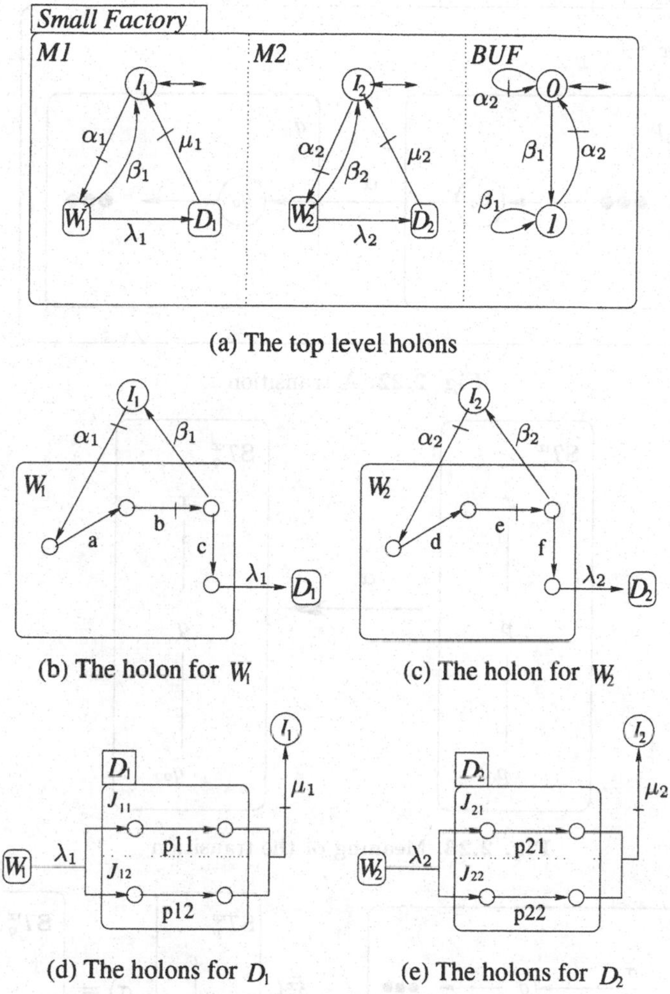

Fig. 2.21. Small Factory III

Let $\sigma \in \Sigma_I^x$. Then construct $\mathbf{ST}^x_q = \bar{\delta}_I^x(\mathbf{ST}^x_p, \sigma)$ based on the following four cases. [15]

1. As illustrated in Figure 2.24, both the source state p and target state q are simple states. Transition σ moves the system from p to q. We get the resulting sub-state-tree $\mathbf{ST}^x_q$ from $\delta_I^x(p, \sigma) = q$.
2. As illustrated in Figure 2.25, the source state p is a simple state, but the target state q is a superstate. In (a), q is an OR superstate. The transition σ moves the system from p to q_o. We get the resulting sub-state-tree $\mathbf{ST}^x_q$ from $\delta_I^x(p, \sigma) = q$ and

[15] In the illustration, we assume our AND superstate has two children; the extension to n children should be clear.

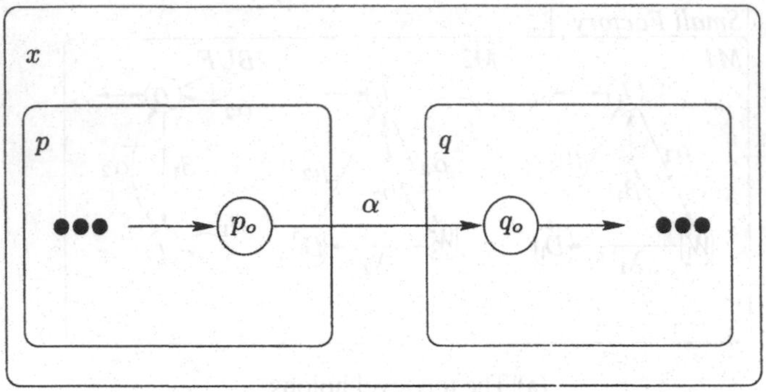

Fig. 2.22. A transition

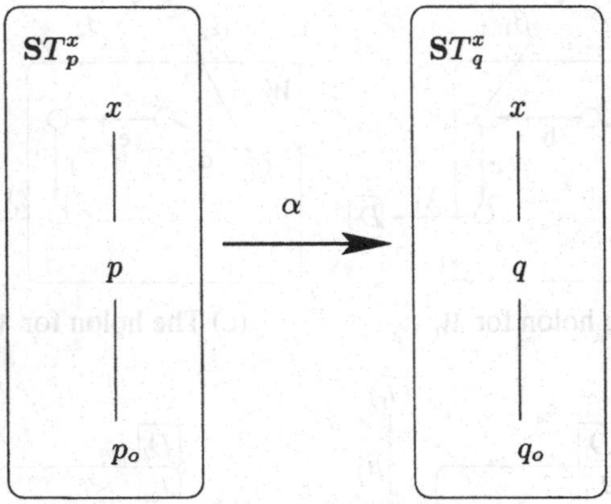

Fig. 2.23. Meaning of the transition

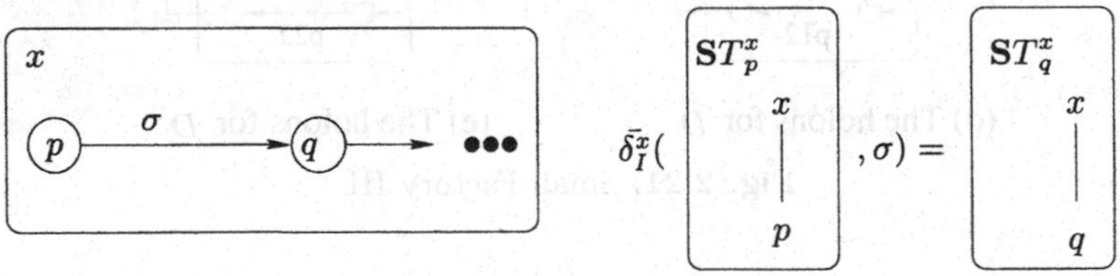

Fig. 2.24. Definition of $\bar{\delta}_I$: case 1

$\delta^q_{BI}(p, \sigma) = q_o$. It includes two steps. First we know that the resulting tree will have superstate q from the high level transition function δ^x_I, and then we know that it must have q_o from the low level transition function δ^q_{BI}. In (b), q is an AND superstate. The transition σ moves the system from p to q_{1o} and q_{2o} at the same time. Again the resulting sub-state-tree $\mathbf{ST}^x_q$ is decided by combining $\delta^x_I(p, \sigma) = q$, $\delta^{q_1}_{BI}(p, \sigma) = q_{1o}$ and $\delta^{q_2}_{BI}(p, \sigma) = q_{2o}$.

3. As illustrated in Figure 2.26, the source state p is a superstate and the target state q is a simple state.

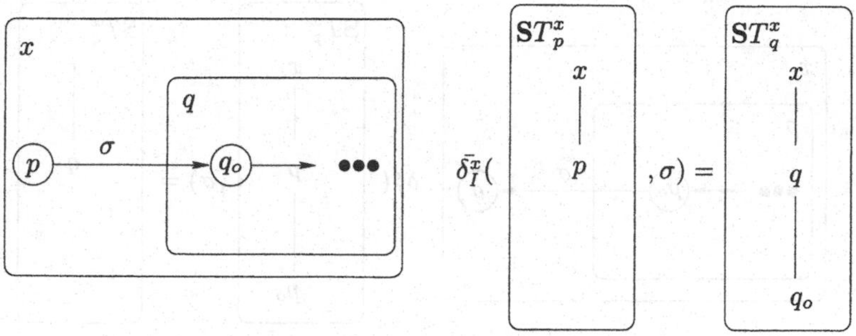

(a) the target state is OR superstate

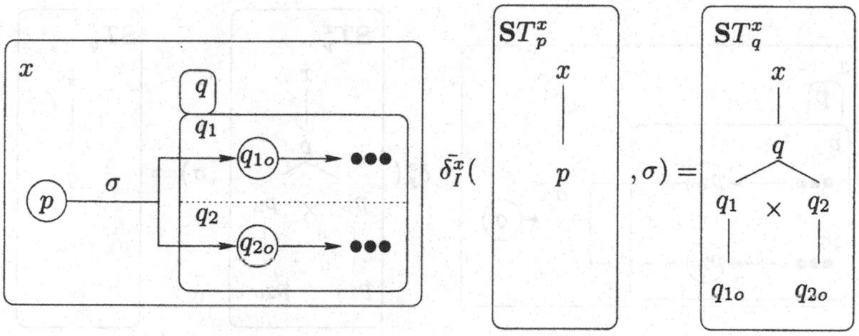

(b) the target state is AND superstate

Fig. 2.25. Definition of $\bar{\delta}_I$: case 2

In (a), p is an OR superstate. The transition σ moves the system from p_o to q. We get the resulting sub-state-tree $\mathbf{ST}^x_q$ from $\delta^x_I(p,\sigma) = q$ and $\delta^p_{BO}(p_o, \sigma) = q$.
In (b), p is an AND superstate. The transition σ moves the system from p_{1o} and p_{2o} to q at the same time. Again the resulting sub-state-tree $\mathbf{ST}^x_q$ is decided by combining $\delta^x_I(p,\sigma) = q$, $\delta^{p_1}_{BO}(p_{1o}, \sigma) = q$ and $\delta^{p_2}_{BO}(p_{2o}, \sigma) = q$.

4. As illustrated in Figure 2.27, both the source state p and target state q are superstates.

◊

Like the δ function, the map $\bar{\delta}^x_I$ is partial; we write $\bar{\delta}^x_I(\mathbf{ST}^x_1, \sigma)!$ if it is defined.

In order to define $\Delta(\mathbf{ST}_1, \sigma)$, the simplest way is to first define a *largest eligible state tree* $\text{Elig}_{\mathbf{G}}(\sigma) : \Sigma \to \mathcal{ST}(\mathbf{ST})$ of σ where σ is allowed to happen, then let $\mathbf{ST}_2 := \mathbf{ST}_1 \wedge \text{Elig}_{\mathbf{G}}(\sigma)$ and replace all the (child) sub-state-trees $\mathbf{ST}^x_2$ by $\bar{\delta}^x_I(\mathbf{ST}^x_2, \sigma)$ on $\mathbf{ST}_2$, whenever $\bar{\delta}^x_I(\mathbf{ST}^x_2, \sigma)!$.

Definition 2.18 [$\text{Elig}_{\mathbf{G}}(\sigma)$] Let $\mathbf{G} = (\mathbf{ST}, \mathcal{H}, \Sigma, \Delta, \mathbf{ST}_o, \mathcal{ST}_m)$ be a state tree structure with $\mathbf{ST} = (X, x_o, \mathcal{T}, \mathcal{E})$. Let $\sigma \in \Sigma$. $\text{Elig}_{\mathbf{G}}(\sigma) : \Sigma \to \mathcal{ST}(\mathbf{ST})$

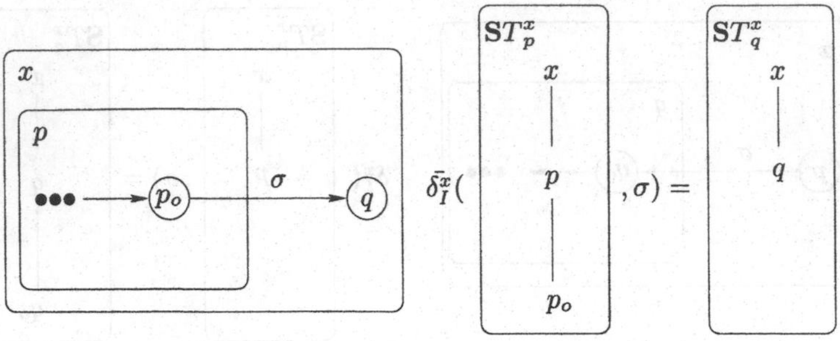

(a) the source state is OR superstate

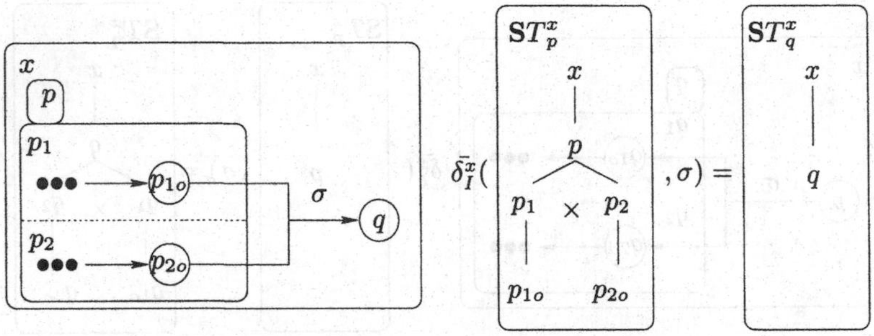

(b) the source state is AND superstate

Fig. 2.26. Definition of $\bar{\delta}_I$: case 3

is the *largest eligible state tree* of σ where σ is allowed to happen. Formally, define $D_\sigma := \{x | \sigma \in \Sigma_I^x\}$ as the set of OR superstates where σ appears in the inner transition structure of their matched holons. Then we have $a \in \mathrm{Elig}_{\mathbf{G}}(\sigma)$ if and only if

1. $(\forall x \in D_\sigma) a | x$, or [16]
2. $(\forall x \in D_\sigma) a \le x$, or
3. $(\exists x \in D_\sigma, T \in \mathcal{B}(\mathbf{ST}^x)) a \in T \;\&\; \bar{\delta}_I^x(T, \sigma)!$. [17]

Say σ is *eligible* for a *nonempty* sub-state-tree $\mathbf{ST}_1$ (of $\mathbf{ST}$) if $\mathbf{ST}_1 \le \mathrm{Elig}_{\mathbf{G}}(\sigma)$.

◇

Remarks

1. A simple example is shown in Figure 2.28. This example will also be used later to demonstrate the computation of Δ.

[16] That is, a and x are parallel.
[17] $\mathbf{ST}^x$ is the child state tree of $\mathbf{ST}$, rooted by x.

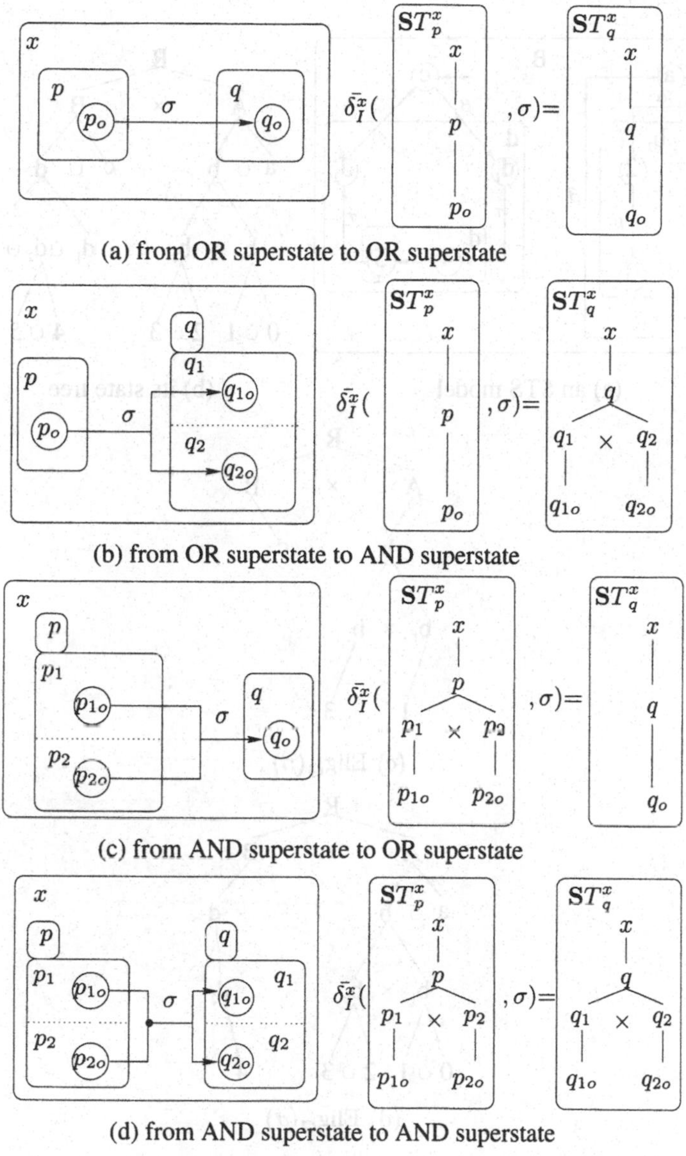

x
p
q
σ
$\delta_I^{\bar{x}}($
ST_p^x
ST_q^x
$,\sigma)=$
(a) from OR superstate to OR superstate
(b) from OR superstate to AND superstate
(c) from AND superstate to OR superstate
(d) from AND superstate to AND superstate

Fig. 2.27. Definition of $\bar{\delta}_I$: case 4

2. In (c) of Figure 2.28, $\text{Elig}_{\mathbf{G}}(\beta)$ is the largest eligible state tree of β. It captures the property that β can occur whenever the system is at states $1, 3$ and c simultaneously.
   Here $D_\beta = \{A, B\}$. So all states except the root state R on $\text{Elig}_{\mathbf{G}}(\beta)$ are given by the item 3 of the above definition.
3. In (d) of Figure 2.28, $\text{Elig}_{\mathbf{G}}(\sigma)$ is the largest eligible state tree of σ. Notice that σ is a local event of the holon matched to d_2. So $\text{Elig}_{\mathbf{G}}(\sigma)$ captures

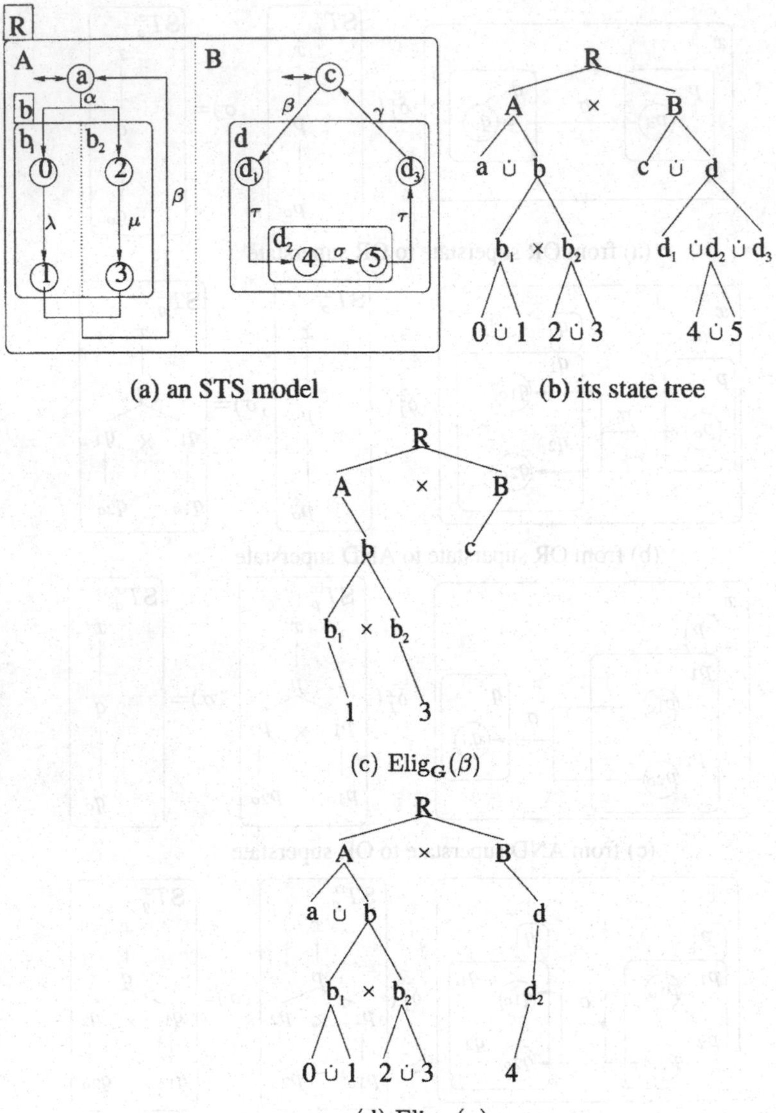

(a) an STS model (b) its state tree

(c) $\mathrm{Elig}_{\mathbf{G}}(\beta)$

(d) $\mathrm{Elig}_{\mathbf{G}}(\sigma)$

Fig. 2.28. Largest eligible state tree of an event

the property that the eligibility of σ has nothing to do with what states the system is staying at, inside $\mathcal{E}^*(A)$.

Here $D_\sigma = \{d_2\}$. The states under A of $\mathrm{Elig}_{\mathbf{G}}(\sigma)$ are given by the item 1 of the above definition, the states in $\{R, B, d\}$ by item 2, and the states in $\{d_2, 4\}$ by item 3.

Definition 2.19 [replace_source$_{\mathbf{G},\sigma}$] Let $\mathbf{G} = (\mathbf{ST}, \mathcal{H}, \Sigma, \Delta, \mathbf{ST}_o, \mathcal{ST}_m)$ be a state tree structure with $\mathbf{ST} = (X, x_o, \mathcal{T}, \mathcal{E})$. Let $\sigma \in \Sigma$. The function replace_source$_{\mathbf{G},\sigma}$: $\mathcal{ST}(\text{Elig}_{\mathbf{G}}(\sigma)) \rightarrow \mathcal{ST}(\mathbf{ST})$ maps a sub-state-tree of $\text{Elig}_{\mathbf{G}}(\sigma)$ to another sub-state-tree of $\mathbf{ST}$ by simply replacing all (child) sub-state-trees T by $\delta_{\bar{I}}^{\bar{x}}(T,\sigma)$ if $\delta_{\bar{I}}^{\bar{x}}(T,\sigma)!$. Formally, define $D_\sigma := \{x | \sigma \in \Sigma_{\bar{I}}^{x}\}$ as the set of OR superstates where σ appears in the inner transition structure of their matched holons. Denote $\mathbf{ST}_2 :=$ replace_source$_{\mathbf{G},\sigma}(\mathbf{ST}_1)$. Then we have $b \in \mathbf{ST}_2$ if and only if

1. $(\forall x \in D_\sigma) b \in \mathbf{ST}_1 \;\&\; b|x$, or [18]
2. $(\forall x \in D_\sigma) b \in \mathbf{ST}_1 \;\&\; b \le x$, or [19]
3. $(\exists x \in D_\sigma, T \in \mathcal{B}(\mathbf{ST}_{\bar{1}}^{x}), T' \in \mathcal{B}(\mathbf{ST}^{x}))\; b \in T' \;\&\; T' = \delta_{\bar{I}}^{\bar{x}}(T,\sigma)$. [20]

◇

Remarks

1. A simple example is shown in Figure 2.29.
2. In (c) of Figure 2.29, the root state R is copied to the resulting sub-state-tree by item 2 of the above definition. All other states are replaced by using item 3.
3. In (d) of Figure 2.29, those states in $\mathcal{E}^*(A)$ and the states in $\{R, B, d\}$ are copied to the resulting sub-state-tree according to item 1 and 2, respectively. Also the states in $\{d_2, 4\}$ are replaced by $\{d_2, 5\}$ according to item 3 of Definition 2.19.

Now we can define Δ by $\text{Elig}_{\mathbf{G}}(\sigma)$ and replace_source$_{\mathbf{G},\sigma}$.

Definition 2.20 [Δ] Let $\mathbf{G} = (\mathbf{ST}, \mathcal{H}, \Sigma, \Delta, \mathbf{ST}_o, \mathcal{ST}_m)$ be a state tree structure with $\mathbf{ST} = (X, x_o, \mathcal{T}, \mathcal{E})$. $\Delta : \mathcal{ST}(\mathbf{ST}) \times \Sigma \rightarrow \mathcal{ST}(\mathbf{ST})$ maps a sub-state-tree of $\mathbf{ST}$ into another sub-state-tree (of $\mathbf{ST}$). Let $T \in \mathcal{ST}(\mathbf{ST})$ and $\sigma \in \Sigma$. Define the total function

$$\Delta(T,\sigma) := \text{replace\_source}_{\mathbf{G},\sigma}(T \wedge \text{Elig}_{\mathbf{G}}(\sigma)).$$

◇

Remarks

1. $\Delta(T,\sigma)$ is a sub-state-tree of $\mathbf{ST}$, according to the definition.
2. Δ maps a basic state tree to another basic state tree or $\emptyset$, the empty state tree.
3. A simple example is shown in Figure 2.30. The resulting state trees are given by the above definition. Notice that in (c), states 0 and 2 are deleted after $(\mathbf{ST}_\beta \wedge \text{Elig}_{\mathbf{G}}(\beta))$, while in (d), state c is deleted after $(\mathbf{ST}_\sigma \wedge \text{Elig}_{\mathbf{G}}(\sigma))$.

---

[18] That is, copy b from $\mathbf{ST}_1$ to $\mathbf{ST}_2$ if b and x are parallel.

[19] That is, copy b from $\mathbf{ST}_1$ to $\mathbf{ST}_2$ if b is an ancestor of all states in D_σ.

[20] That is, replace all basic (child) state trees T by $\delta_{\bar{I}}^{\bar{x}}(T,\sigma)$ if $\delta_{\bar{I}}^{\bar{x}}(T,\sigma)!$.

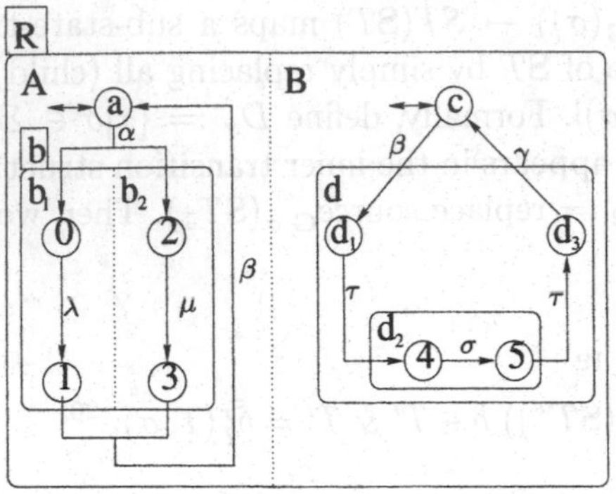

(a) an STS model

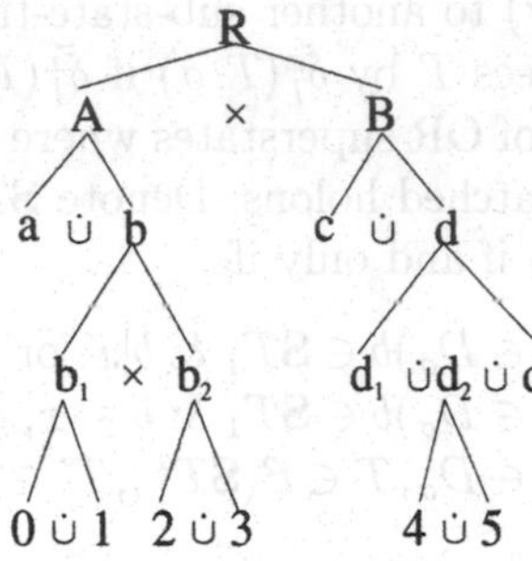

(b) its state tree

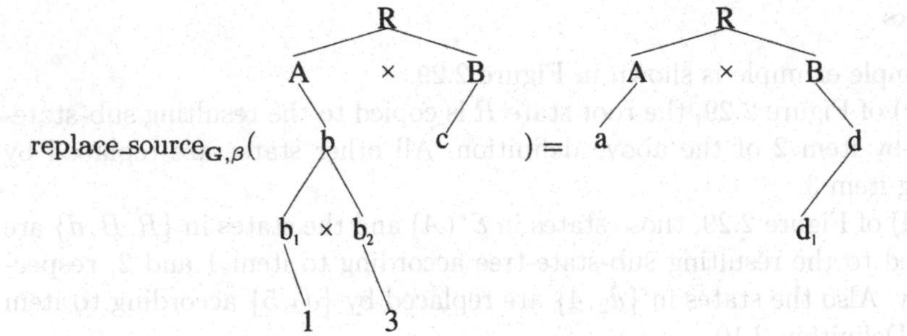

(c) replace_source$_{\mathbf{G},\beta}(\mathbf{ST}_\beta)$

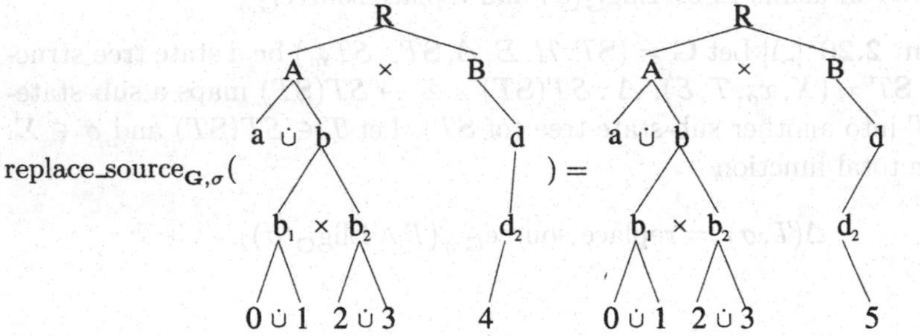

(d) replace_source$_{\mathbf{G},\sigma}(\mathbf{ST}_\sigma)$

Fig. 2.29. Example for replace_source

4. The Δ function is total. $\Delta(T, \sigma)$ returns the empty state tree if $T \wedge \text{Elig}_{\mathbf{G}}(\sigma) = \emptyset$.
5. As we just need to visit each state on T at most once for computing $T \wedge \text{Elig}_{\mathbf{G}}(\sigma)$ and replace_source$_{\mathbf{G},\sigma}(\cdot)$, respectively, the computational complexity of Δ is $O(\|T\|)$, where $\|T\|$ is the number of states (i.e., nodes) on T.

   Notice that the size of $\mathcal{B}(T)$ can be enormous, corresponding to a large set of RW states. So symbolically computing the set of target RW states

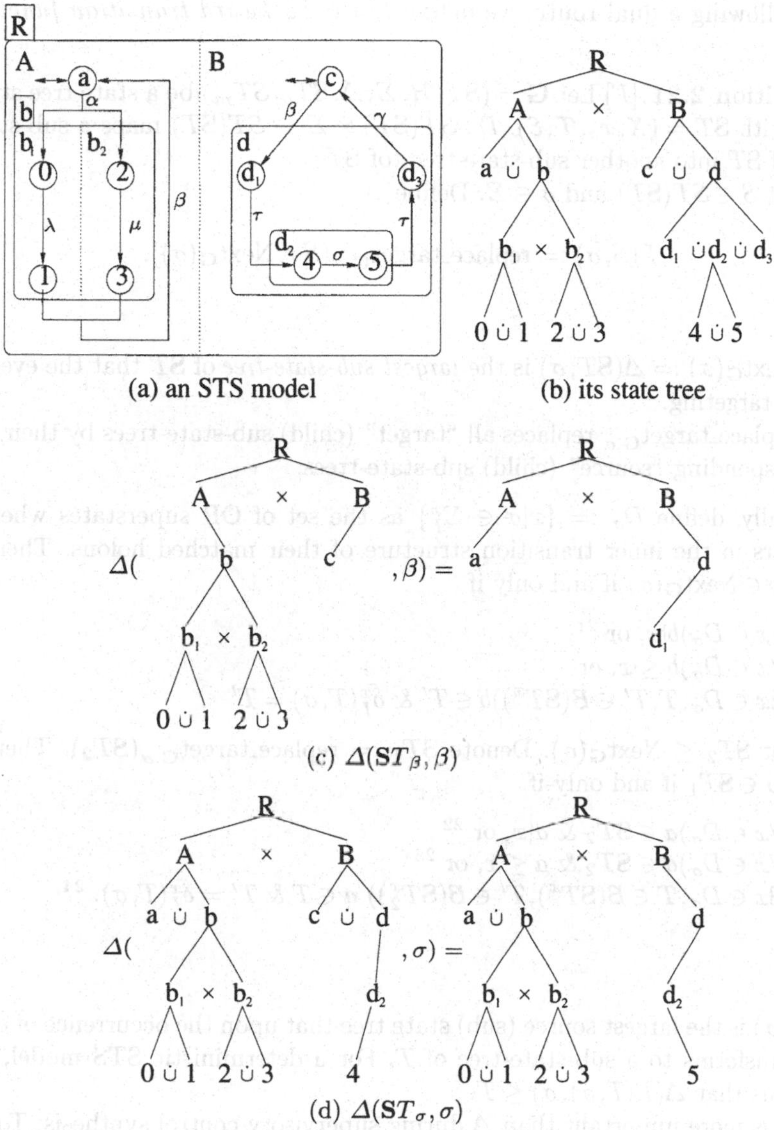

Fig. 2.30. Example for the Δ function

using Δ function has a significant advantage. That is, each call of the Δ function on a sub-state-tree T is equivalent to $|\mathcal{B}(T)|$ (or count(T)) calls of the δ function.

Δ can be used to *simulate* the system behavior, which computes the next (basic) state tree from the given current (basic) state tree. For *synthesis*, we also need a function Γ, which should perform exactly the inverse of Δ, i.e., compute the source state tree from the given target state tree.

Following a dual route, we define Γ, the *backward transition function* of **G**.

Definition 2.21 [Γ] Let $\mathbf{G} = (\mathbf{ST}, \mathcal{H}, \Sigma, \Delta, \mathbf{ST}_o, \mathcal{ST}_m)$ be a state tree structure with $\mathbf{ST} = (X, x_o, \mathcal{T}, \mathcal{E})$. $\Gamma : \mathcal{ST}(\mathbf{ST}) \times \Sigma \to \mathcal{ST}(\mathbf{ST})$ maps a sub-state-tree of $\mathbf{ST}$ into another sub-state-tree (of $\mathbf{ST}$).

Let $S \in \mathcal{ST}(\mathbf{ST})$ and $\sigma \in \Sigma$. Define

$$\Gamma(S, \sigma) := \text{replace\_target}_{\mathbf{G},\sigma}(S \wedge \text{Next}_{\mathbf{G}}(\sigma)).$$

where

1. $\text{Next}_{\mathbf{G}}(\sigma) := \Delta(\mathbf{ST}, \sigma)$ is the *largest sub-state-tree* of $\mathbf{ST}$ that the event σ is targeting.
2. $\text{replace\_target}_{\mathbf{G},\sigma}$ replaces all "target" (child) sub-state-trees by their corresponding "source" (child) sub-state-trees.

Formally, define $D_\sigma := \{x | \sigma \in \Sigma_I^x\}$ as the set of OR superstates where σ appears in the inner transition structure of their matched holons. Then we have $b \in \text{Next}_{\mathbf{G}}(\sigma)$ if and only if

1. $(\forall x \in D_\sigma) b | x$, or [21]
2. $(\forall x \in D_\sigma) b \le x$, or
3. $(\exists x \in D_\sigma, T, T' \in \mathcal{B}(\mathbf{ST}^x)) b \in T'$ & $\bar{\delta}_I^x(T, \sigma) = T'$

Let $\mathbf{ST}_2 \le \text{Next}_{\mathbf{G}}(\sigma)$. Denote $\mathbf{ST}_1 := \text{replace\_target}_{\mathbf{G},\sigma}(\mathbf{ST}_2)$. Then we have $a \in \mathbf{ST}_1$ if and only if

1. $(\forall x \in D_\sigma) a \in \mathbf{ST}_2$ & $a | x$, or [22]
2. $(\forall x \in D_\sigma) a \in \mathbf{ST}_2$ & $a \le x$, or [23]
3. $(\exists x \in D_\sigma, T \in \mathcal{B}(\mathbf{ST}^x), T' \in \mathcal{B}(\mathbf{ST}_2^x))$ $a \in T$ & $T' = \bar{\delta}_I^x(T, \sigma)$. [24]

◊

$\Gamma(T, \sigma)$ is the largest source (sub) state tree that upon the occurrence of event σ, transforms to a sub-state-tree of T. For a deterministic STS model, it is obvious that $\Delta(\Gamma(T, \sigma), \sigma) \le T$.

Γ is more important than Δ during supervisory control synthesis. To this end we will introduce a symbolic computation method for Γ in chapter 4.

[21] That is, b and x are parallel.
[22] That is, copy a from $\mathbf{ST}_2$ to $\mathbf{ST}_1$ if a and x are parallel.
[23] That is, copy a from $\mathbf{ST}_2$ to $\mathbf{ST}_1$ if a is an ancestor of all states in D_σ.
[24] That is, replace all (child) sub-state-trees T' by T if $\bar{\delta}_I^x(T, \sigma) = T'$.

Discussion on the Soundness of Δ

In the original RW framework, the transition structure is defined over a *flat state* set. Then given a set A of source flat states and an event σ, to compute its target set B of flat states we need to check every member $p \in A$ and if $\delta(p,\sigma)!$, put $\delta(p,\sigma)$ in B. This extensional approach is computationally very expensive for complex systems. In our STS setting, if the given set A can be represented by a sub-state-tree S, i.e., $A = \mathcal{B}(S)$, then theoretically we require $\mathcal{B}(\Delta(S,\sigma)) = B$ in order for our intensional definition of Δ to be *sound.*

From the definition of Δ, we need both $\text{Elig}_{\mathbf{G}}$ and $\text{replace\_source}_{\mathbf{G},\sigma}$ to be *sound.*

Let us investigate the soundness of $\text{Elig}_{\mathbf{G}}$ first. In order to do this, we need to define the correct eligible set of basic state trees (i.e., the correct set of flat states in the original RW framework) by extensionality.

Definition 2.22 [RW-eligible ,$E_{\mathbf{G},\sigma}$] Let $\mathbf{G} = (\mathbf{ST}, \mathcal{H}, \Sigma, \Delta, \mathbf{ST}_o, ST_m)$ be a state tree structure with $\mathbf{ST} = (X, x_o, \mathcal{T}, \mathcal{E})$. Let $\sigma \in \Sigma$. Let $D_\sigma := \{x | \sigma \in \Sigma_I^x\}$ be the set of OR superstates where σ appears in the inner transition structure of their matched holons. Then a basic state tree $b \in \mathcal{B}(\mathbf{ST})$ is *RW-eligible for* σ if

$$(\forall x \in D_\sigma) x \in b \;\&\; \delta_I^x(b^x, \sigma)!.$$

Here b^x is the child state tree of b, rooted by x. Write the set

$$E_{\mathbf{G},\sigma} = \{b | b \text{ is RW-eligible for } \sigma\}$$

as the set of all RW-eligible basic state trees for σ.

◊

In Figure 2.31, an STS model is given in (a) and its state tree is given in (b). We list all of the elements of $E_{\mathbf{G},\sigma}$ in (c).

It is of interest to find the relationship between the intensional definition of $\text{Elig}_{\mathbf{G}}(\sigma)$ and the extensional definition of $E_{\mathbf{G},\sigma}$.

Lemma 2.10 $\text{Elig}_{\mathbf{G}}(\sigma) = \bigvee_{b \in E_{\mathbf{G},\sigma}} b.$

Proof. 1. Show that $(\bigvee_{b \in E_{\mathbf{G},\sigma}} b) \leq \text{Elig}_{\mathbf{G}}(\sigma)$.
Let $b \in E_{\mathbf{G},\sigma}$. Then $b \leq \text{Elig}_{\mathbf{G}}(\sigma)$ because from the definition, every state on b must be on $\text{Elig}_{\mathbf{G}}(\sigma)$ too. Therefore, $(\bigvee_{b \in E_{\mathbf{G},\sigma}} b) \leq \text{Elig}_{\mathbf{G}}(\sigma)$ as required.

2. Show that $\text{Elig}_{\mathbf{G}}(\sigma) \leq \bigvee_{b \in E_{\mathbf{G},\sigma}} b$.
We prove it by contradiction. Let y be a leaf state that is on the tree $\text{Elig}_{\mathbf{G}}(\sigma)$ but not on the tree $\bigvee_{b \in E_{\mathbf{G},\sigma}} b$, i.e., y should not be on any basic state tree $b \in E_{\mathbf{G},\sigma}$. Let $D_\sigma := \{x | \sigma \in \Sigma_I^x\}$ be the set of OR superstates where σ appears in the inner transition structure of their matched holons. If $\{y\} | D_\sigma$ [25], then y must be on some basic state tree $b \in E_{\mathbf{G},\sigma}$, because b

[25] Refer to the definition of | on page 19.

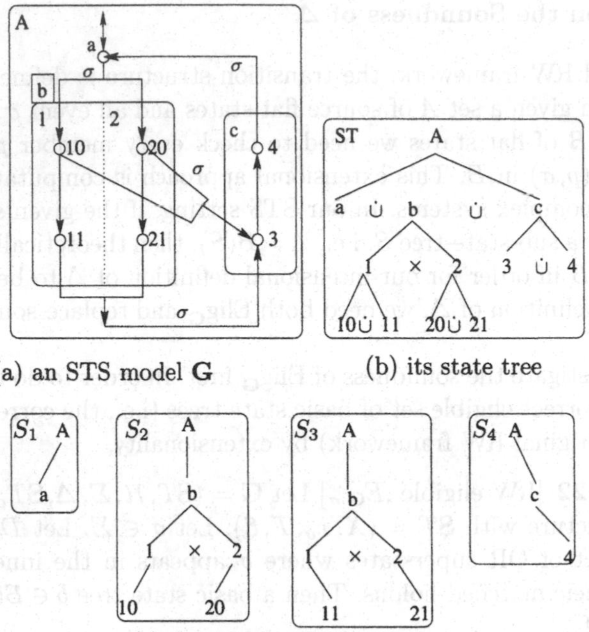

(a) an STS model **G** (b) its state tree

(c) RW-eligible basic state trees

Fig. 2.31. Example: $E_{\mathbf{G},\sigma}$

only needs to satisfy the conditions for the descendants of the states in D_σ. So there must exist $x \in D_\sigma$ such that $x < y$. Since y should not be on any basic state tree $b \in E_{\mathbf{G},\sigma}$, $(\forall T \in \mathcal{B}(\mathbf{ST}^x))y \in T \Rightarrow (\delta_I^x(T,\sigma)$ not defined). Then y should not be on $\text{Elig}_{\mathbf{G}}(\sigma)$ too, from the definition of $\text{Elig}_{\mathbf{G}}(\sigma)$, a contradiction. So every leaf state and its ancestors on $\text{Elig}_{\mathbf{G}}(\sigma)$ must be on $\bigvee_{b\in E_{\mathbf{G},\sigma}} b$, i.e., every state on $\text{Elig}_{\mathbf{G}}(\sigma)$ must be on $\bigvee_{b\in E_{\mathbf{G},\sigma}} b$. Therefore, $\text{Elig}_{\mathbf{G}}(\sigma) \leq \bigvee_{b\in E_{\mathbf{G},\sigma}} b$.

From Lemma 2.9, $\mathcal{B}(\mathbf{ST}_1)\cup\mathcal{B}(\mathbf{ST}_2) \subseteq \mathcal{B}(\mathbf{ST}_1\vee\mathbf{ST}_2)$. Then $E_{\mathbf{G},\sigma} \subseteq \mathcal{B}(\text{Elig}_{\mathbf{G}}(\sigma))$. However, it may not always be the case that $E_{\mathbf{G},\sigma} = \mathcal{B}(\text{Elig}_{\mathbf{G}}(\sigma))$. For the example in Figure 2.31, $\text{Elig}_{\mathbf{G}}(\sigma) = S_1 \vee S_2 \vee S_3 \vee S_4$ is given in Figure 2.32. Unfortunately, $E_{\mathbf{G},\sigma} \subset \mathcal{B}(\text{Elig}_{\mathbf{G}}(\sigma))$.

Definition 2.23 [sound $\text{Elig}_{\mathbf{G}}$] $\text{Elig}_{\mathbf{G}}$ is *sound* if for any given event σ,

$$\mathcal{B}(\text{Elig}_{\mathbf{G}}(\sigma)) = E_{\mathbf{G},\sigma}.$$

◇

For the example in Figure 2.31, it is easy to find that $\text{Elig}_{\mathbf{G}}$ is not sound. This is because the AND superstate in the example has two outgoing transitions with the same event label σ. Generally, let $D_\sigma := \{x|\sigma \in \Sigma_I^x\}$ be the set

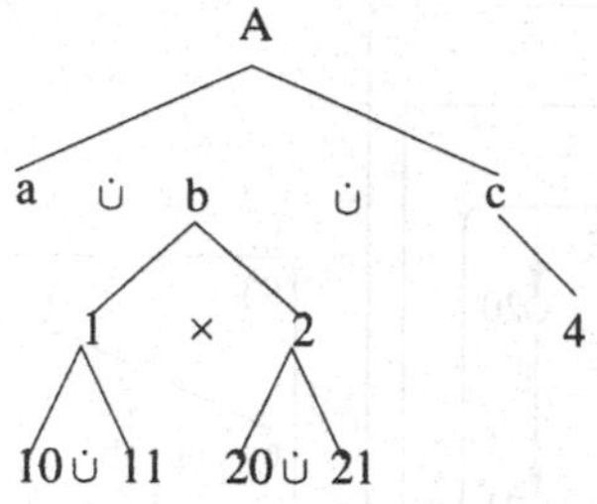

Fig. 2.32. $\text{Elig}_{\mathbf{G}}(\sigma)$

of OR superstates where σ appears in the inner transition structure of their matched holons. Thanks to local coupling, $\text{Elig}_{\mathbf{G}}$ is sound if for any $x \in D_\sigma$, $\text{Elig}_{\mathbf{G}^x}$ is sound, where $\mathbf{G}^x$ is the child STS of $\mathbf{G}$, rooted by x. So we can give a sufficient condition which ensures the soundness of $\text{Elig}_{\mathbf{G}}$: *every outgoing boundary transition of the holon matched to an AND component must have a unique event label.* This is essentially not very restrictive, since we can usually relabel a transition by a new event during the modelling stage. For the example in Figure 2.31, if we relabel one of the outgoing boundary transitions in holon 1 and 2 by a new event, say β, then $\text{Elig}_{\mathbf{G}}$ is sound.

Following a similar route, we investigate the soundness of $\text{replace\_source}_{\mathbf{G},\sigma}$.

Definition 2.24 Assume that $\text{Elig}_{\mathbf{G}}$ is sound. For any $S \in \mathcal{ST}(\text{Elig}_{\mathbf{G}}(\sigma))$, define

$$R_{\mathbf{G},\sigma}(S) := \bigcup_{b \in \mathcal{B}(S)} \mathcal{B}(\text{replace\_source}_{\mathbf{G},\sigma}(b)).$$

Then $\text{replace\_source}_{\mathbf{G},\sigma}$ is *sound* if for any $S \in \mathcal{ST}(\text{Elig}_{\mathbf{G}}(\sigma))$,

$$\mathcal{B}(\text{replace\_source}_{\mathbf{G},\sigma}(S)) = R_{\mathbf{G},\sigma}(S).$$

$\diamond$

For the example in Figure 2.33, it is easy to find that $\text{replace\_source}_{\mathbf{G},\sigma}$ is not sound, because $R_{\mathbf{G},\sigma}(S_3) = \{S_1', S_2'\} \subset \mathcal{B}(\text{replace\_source}_{\mathbf{G},\sigma}(S_3))$. The reason is that the AND superstate has two incoming transitions with the same event label σ. So for any given event σ, we can give a sufficient condition that ensures the soundness of $\text{replace\_source}_{\mathbf{G},\sigma}$: *every incoming boundary transition of the holon matched to an AND component must have a unique event label.* This condition is also not very restrictive because we can relabel transitions by new events.

To compute $\Delta(T, \sigma)$, essentially we need to

1. compute $K_1 := \text{Elig}_{\mathbf{G}}(\sigma)$, then
2. compute $K_2 := T \wedge K_1$, and finally
3. compute $\text{replace\_source}_{\mathbf{G},\sigma}(K_2)$.

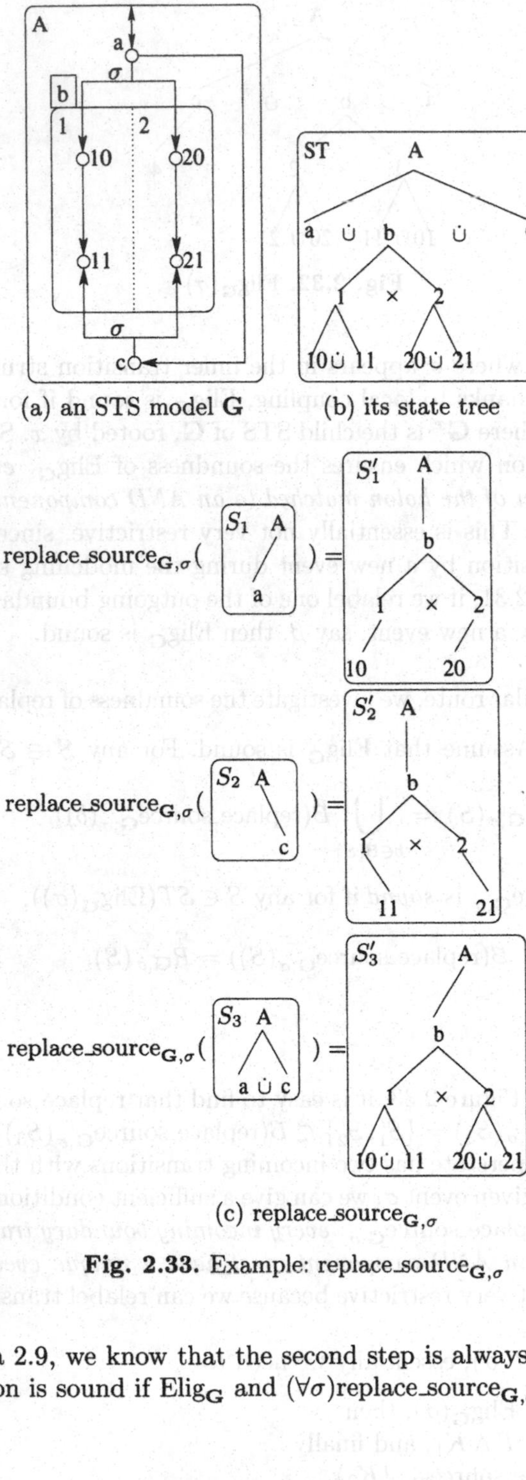

Fig. 2.33. Example: replace_source$_{\mathbf{G},\sigma}$

From Lemma 2.9, we know that the second step is always sound. Therefore, the Δ function is sound if $\text{Elig}_\mathbf{G}$ and $(\forall\sigma)\text{replace\_source}_{\mathbf{G},\sigma}$ are sound.

2.4 Summary

The concept of state tree is introduced to give structure to the state space. A sub-state-tree $\mathbf{S}T_1$ is the "symbol" of $\mathcal{B}(\mathbf{S}T_1)$, the set of basic sub-state-trees of $\mathbf{S}T_1$, i.e., $\mathbf{S}T_1$ tells us the characteristics of the set $\mathcal{B}(\mathbf{S}T_1)$. We also proved that $(\mathcal{S}T(\mathbf{S}T), \leq)$ is a lattice. This property underlies the definition of the global transition function Δ.

In a state tree structure (STS), holons are used to define local transition structures; the dynamics of the whole system is given by the Δ function. This setup is important because we cannot afford to store the whole transition graph for complex systems. However, we can store the dynamics of each holon in the STS and write an algorithm to *compute* Δ. For this reason, the Δ function has a significant computational advantage over the δ function.

3

Nonblocking Supervisory Control of State Tree Structures

In the previous chapter, we explained how to model a system using state tree structures (STS) and described its behavior by the Δ function. In this chapter, we will move on to deal with the control of STS. The setup of this chapter is the following. First, we compare the state trees of our STS with predicates in section 3.1. Then we handle controllability and nonblocking in section 3.2 and 3.3, respectively. In outline, this chapter follows a similar approach to that of chapter 7 of [Won04]. A new feature is the introduction of *weak controllability*. With that, the onerous requirement of reachability checking during the synthesis stage can be dropped, with no change in the resulting controlled behavior, but with a gain in computational efficiency for large systems.

The primary objective of this chapter is to prove the existence of optimal (maximally permissive) state feedback controllers. The actual computation will be given in the next chapter, where efficiency is the focus.

3.1 State Trees and Predicates

It is well known that for each state subset in RW framework, there exists a unique predicate defined over the state space such that every state of the subset satisfies the predicate (see chapter 7 of [Won04]). We find that state trees and predicates also have a close relationship.

Let $\mathbf{G} = (ST, \mathcal{H}, \Sigma, \Delta, \mathbf{ST}_o, ST_m)$ be a state tree structure with $ST = (X, x_o, \mathcal{T}, \mathcal{E})$. A predicate P defined on $\mathcal{B}(\mathbf{ST})$ (or simply on $\mathbf{ST}$) is a function $P : \mathcal{B}(\mathbf{ST}) \to \{0,1\}$. [1] Also a predicate can be *identified* by a set of basic state trees, say B_P, such that

$$B_P := \{b \in \mathcal{B}(\mathbf{ST}) | P(b) = 1\} \subseteq \mathcal{B}(\mathbf{ST}).$$

[1] In Chapter 7 of [Won04], P is defined on Q such that $P : Q \to \{0,1\}$. This is consistent with our definition as a basic state tree in our STS setting is equivalent to a flat state in RW framework.

If there exists $\mathbf{ST}_1 \leq \mathbf{ST}$ such that $B_P = \mathcal{B}(\mathbf{ST}_1)$, we say that P is *identified* by $\mathbf{ST}_1$. [2]

Let $b \in \mathcal{B}(\mathbf{ST})$. We say predicate P *holds*, or is *satisfied* for b, i.e., $b \models P$ if and only if $b \in B_P$. Let $\mathbf{ST}_1 \leq \mathbf{ST}$. We say predicate P *holds*, or is *satisfied* for $\mathbf{ST}_1$, i.e., $\mathbf{ST}_1 \models P$ if and only if $\mathcal{B}(\mathbf{ST}_1) \subseteq B_P$. Write *Pred*(**ST**) as the set of all predicates defined on **ST**. Then for convenience, write $(\forall P \in Pred(\mathbf{ST}))\emptyset \models P$, where $\emptyset$ is the empty state tree. [3]

Propositional logic operators are defined in a straightforward fashion using standard notation:

$$\begin{array}{ll} (\neg P)(b) = 1 & \text{iff } P(b) = 0; \\ (P_1 \wedge P_2)(b) = 1 & \text{iff } P_1(b) = 1 \text{ and } P_2(b) = 1; \\ (P_1 \vee P_2)(b) = 1 & \text{iff } P_1(b) = 1 \text{ or } P_2(b) = 1. \end{array}$$

We also can define a partial order on *Pred*(**ST**) from subset containment:

$$P_1 \preceq P_2 \text{ iff } P_1 \wedge P_2 = P_1.$$

That is, P_1 precedes P_2, or is stronger than P_2, if and only if $(\forall b) b \models P_1 \Rightarrow b \models P_2$. Say P_1 is a *subpredicate* of P_2. Under the definition, $(Pred(\mathbf{ST}), \preceq)$ is a lattice. The top element $\top$ is defined by $B_\top = \mathcal{B}(\mathbf{ST})$ and the bottom element $\bot$ is defined by $B_\bot = \emptyset$, the empty set of basic (sub) state trees.

In the STS **G**, the *initial predicate* P_o is defined by $B_{P_o} = \mathcal{B}(\mathbf{ST}_o)$ [4], and the *marker predicate* P_m by $B_{P_m} = \bigcup_{T \in \mathcal{ST}_m} \mathcal{B}(T)$. Now we can rewrite the plant STS by $\mathbf{G} = (\mathbf{ST}, \mathcal{H}, \Sigma, \Delta, P_o, P_m)$, where $\mathbf{ST}_o$ and $\mathcal{ST}_m$ are replaced by their predicate counterparts.

3.2 Controllability and State Feedback Control

Let $\mathbf{G} = (\mathbf{ST}, \mathcal{H}, \Sigma, \Delta, P_o, P_m)$ be a state tree structure. The *reachability predicate* $R(\mathbf{G}, P)$ is defined to designate all the basic state trees that can be reached from initial basic state trees via state trees satisfying P:

1. $P \wedge P_o = false \Rightarrow R(\mathbf{G}, P) = false$;
2. $(b_o \models P \wedge P_o) \Rightarrow (b_o \models R(\mathbf{G}, P))$;
3. $b \models R(\mathbf{G}, P)$ & $\sigma \in \Sigma$ & $\Delta(b, \sigma) \neq \emptyset$ & $\Delta(b, \sigma) \models P \Rightarrow \Delta(b, \sigma) \models R(\mathbf{G}, P)$;
4. No other basic state trees satisfy $R(\mathbf{G}, P)$.

With **G** fixed, $R(\mathbf{G}, \cdot) : Pred(\mathbf{ST}) \rightarrow Pred(\mathbf{ST})$ is a *predicate transformer*. Clearly $R(\mathbf{G}, \cdot)$ is monotone: $P \preceq Q$ implies $R(\mathbf{G}, P) \preceq R(\mathbf{G}, Q)$.

---

[2] In this case, usually P can enjoy a short formula representation.

[3] The left hand side of $\models$ is always a sub-state-tree, whereas the right hand side of $\models$ is always a predicate.

[4] Notice that the size of $\mathcal{B}(\mathbf{ST}_o)$ is greater than 1 in general.

The *weakest liberal precondition* is the predicate transformer $M_\sigma : Pred(\mathbf{ST}) \rightarrow Pred(\mathbf{ST})$ defined as follows:

$$b \models M_\sigma(P) \text{ iff } \Delta(b,\sigma) \models P.$$

Notice that if $\Delta(b,\sigma) = \emptyset$, then for any P, $\Delta(b,\sigma) \models P$. This definition agrees with [Won04].

We also let $f : \mathcal{B}(\mathbf{ST}) \rightarrow \Pi$ denote the *state feedback control* (SFBC) for $\mathbf{G}$, where

$$\Pi := \{\Sigma' \subseteq \Sigma | \Sigma_u \subseteq \Sigma'\}.$$

The event σ is *enabled* at b if $\sigma \in f(b)$, and is *disabled* otherwise. For event σ introduce the predicate $f_\sigma : \mathcal{B}(\mathbf{ST}) \rightarrow \{0,1\}$ defined by

$$f_\sigma(b) := 1 \text{ iff } \sigma \in f(b).$$

Thus the SFBC f can be implemented by the set of predicates $\{f_\sigma | \sigma \in \Sigma\}$.

The closed-loop transition function induced by the SFBC f is given by

$$\Delta^f(b,\sigma) := \begin{cases} \Delta(b,\sigma), \text{if } f_\sigma(b) = 1 \\ \emptyset, \quad\quad\quad \text{otherwise} \end{cases}.$$

Unlike Δ, Δ^f need be defined only on basic state trees. We write the controlled STS as $\mathbf{G}^f := (\mathbf{ST}, \mathcal{H}, \Sigma, \Delta^f, P_o^f, P_m)$ with $P_o^f \preceq P_o$ for the closed system supervised by the SFBC f. Notice that in general some initial basic state trees are excluded from $\mathbf{G}^f$. To choose $P_o^f (\preceq P_o)$, the allowable initial basic state trees, is also the responsibility of the synthesizer.

Now we can talk about controllability.

A predicate $P \in Pred(\mathbf{ST})$ is *weakly controllable* [5] wrt. $\mathbf{G}$ if

$$(\forall \sigma \in \Sigma_u) P \preceq M_\sigma(P).$$

The reason we drop the condition $P \preceq R(\mathbf{G}, P)$ imposed in chapter 7 of [Won04] is that the computation of $R(\mathbf{G}, P)$ is expensive [6] and, as will be seen later, unnecessary for the synthesis of SFBC f.

Let's use a simple example to demonstrate our weak controllability. Consider the finite state machine in Figure 3.1. Table 3.1 illustrates the difference between weak controllability and controllability. There are three states in the machine, 0, 1 and 2. In the table we check all possible 2^3 predicates. Every

[5] This is not a fundamental new concept. In chapter 7 of [Won04], a predicate is controllable if it is reachable from the initial state as well as invariant under the occurrence of uncontrollable events. Formally, P is *controllable* if and only if

$$P \preceq R(\mathbf{G}, P) \ \ \&(\forall \sigma \in \Sigma_u) P \preceq M_\sigma(P).$$

[6] The efficient computation of reachable predicates for complex systems is still an open problem.

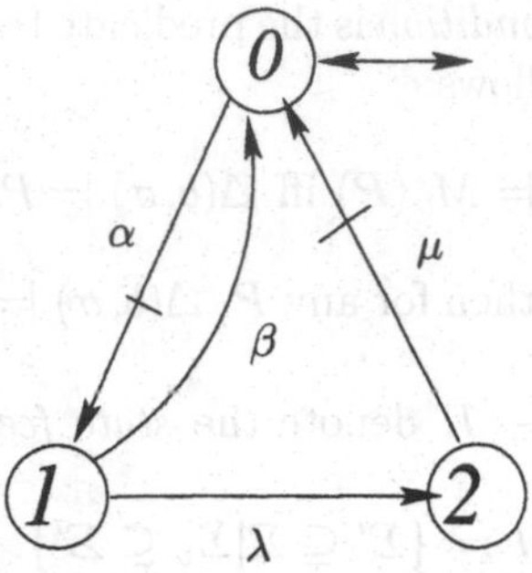

Fig. 3.1. Example: Weak Controllability

Table 3.1. Comparison: weak controllability and controllability

B_P	**Weakly Controllable**	**Controllable**
$\emptyset$	Yes	Yes
$\{0\}$	Yes	Yes
$\{1\}$	No	No
$\{2\}$	Yes	No
$\{0,1\}$	No	No
$\{0,2\}$	Yes	No
$\{1,2\}$	No	No
$\{0,1,2\}$	Yes	Yes

controllable predicate is also weakly controllable. Some predicates, identified by $\{2\}$ and $\{0,2\}$, are only weakly controllable. But their corresponding *reachable subpredicates*, identified by $\emptyset$ and $\{0\}$, respectively, are controllable. This comes from Proposition 3.1.

Proposition 3.1 Let P be a weakly controllable predicate of $\mathbf{G}$. Then $R(\mathbf{G}, P)$ is controllable.

Proof. $R(\mathbf{G},P) \preceq R(\mathbf{G},R(\mathbf{G},P))$ because $R(\mathbf{G},P)$ is already reachable. We just need to show that $(\forall \sigma \in \Sigma_u)R(\mathbf{G},P) \preceq M_\sigma(R(\mathbf{G},P))$.

Assume $b \models R(\mathbf{G},P)$. Of course $b \models P$ as $R(\mathbf{G},P) \preceq P$. Thus, since P is weakly controllable, $(\forall \sigma \in \Sigma_u)\Delta(b,\sigma) \models P$. Then let σ be an uncontrollable event. If $\Delta(b,\sigma) = \emptyset$, $\Delta(b,\sigma) \models R(\mathbf{G},P)$ (any predicate holds for the empty state tree). Otherwise ($\Delta(b,\sigma) \neq \emptyset$), from the definition of $R(\mathbf{G},P)$,

$$\Delta(b,\sigma) \models R(\mathbf{G},P),$$

i.e., $(\forall \sigma \in \Sigma_u)b \models M_\sigma(R(\mathbf{G},P))$, as required.

Given a predicate, it is easier to verify if it is weakly controllable. What one needs is just some local information, as to obtain $M_\sigma(P)$ only the one step transition function Δ is called. Without computing expensive reachable predicate $R(\mathbf{G}, P)$, we gain computational efficiency. But is it adequate just to get a weakly controllable predicate? Theorem 3.1 says so.

Theorem 3.1 Let $P \in Pred(\mathbf{ST})$ and $P \wedge P_o \neq false$. If P is weakly controllable, then there exists a SFBC f for $\mathbf{G}$ such that $R(\mathbf{G}^f, true) = R(\mathbf{G}, P)$, where $\mathbf{G}^f := (\mathbf{ST}, \mathcal{H}, \Sigma, \Delta^f, P_o^f, P_m)$ and $P_o^f = P \wedge P_o$.

Remark: This is one of the most important theorems in this chapter. It guarantees that given a *weakly controllable* predicate P, there *exists* a SFBC f to implement its reachable subpredicate $R(\mathbf{G}, P)$. Therefore, it underlies the synthesis procedure. A similar result is already given in chapter 7 of [Won04]. However, that result requires a *controllable* predicate P, which is computationally more expensive to verify.

Proof. Assume P is weakly controllable. Define the SFBC $f : Q \to \Gamma$ by

$$(\forall \sigma \in \Sigma_c) f_\sigma := M_\sigma(P).$$

1. First we show that $R(\mathbf{G}^f, true) \preceq R(\mathbf{G}, P)$.
Let $b \models R(\mathbf{G}^f, true)$. Since $\Delta^f(b, \sigma) \neq \emptyset$ implies $\Delta(b, \sigma) \neq \emptyset$, we have

$$\Delta(b_i, \sigma_i) = b_{i+1}, \quad i = 0, 1, \dots, k-1,$$

$$f_{\sigma_i}(b_i) = 1, \quad i = 0, 1, \dots, k-1,$$

for some $k > 0, b_o(\models P_o^f), b_1, \dots, b_k(= b) \in \mathcal{B}(\mathbf{ST})$ and $\sigma_o, \sigma_1, \dots, \sigma_{k-1} \in \Sigma$. Since $P_o^f = P_o \wedge P$, $b_o \models P$. We can show that $b_1 \models P$. For if $\sigma_o \in \Sigma_u$, we have $b_o \models M_{\sigma_o}(P)$ as P is weakly controllable, i.e., $b_1 = \Delta(b_o, \sigma_o) \models P$; while if $\sigma_o \in \Sigma_c$, $f_{\sigma_o}(b_o) = 1$, namely $b_o \models M_{\sigma_o}(P)$ and again $b_1 = \Delta(b_o, \sigma_o) \models P$. Repeatedly we can show that for all b_i on this path, $b_i \models P$. Thus $b \models R(\mathbf{G}, P)$.
2. Next we show that $R(\mathbf{G}, P) \preceq R(\mathbf{G}^f, true)$.
Let $b \models R(\mathbf{G}, P)$. Then for some $k \geq 0, b_o(\models P_o), b_1, \dots, b_k(= b) \in \mathcal{B}(\mathbf{ST})$ and $\sigma_o, \sigma_1, \dots, \sigma_{k-1} \in \Sigma$, we have

$$b_i \models P, \quad i = 0, 1, \dots, k,$$

$$\Delta(b_i, \sigma_i) = b_{i+1}, \quad i = 0, 1, \dots, k-1,$$

and therefore

$$b_i \models M_{\sigma_i}(P), \quad i = 0, 1, \dots, k-1.$$

Notice that $P_o^f = P_o \wedge P$. Thus $b_o \models P_o^f$. We can show that $\Delta^f(b_i, \sigma_i) = b_{i+1} \neq \emptyset$, for all $i = 0, 1, \dots, k-1$. For if $\sigma_i \in \Sigma_u$, then $f_{\sigma_i}(b_i) = 1$ because f is SFBC; while if $\sigma_i \in \Sigma_c$, then $f_{\sigma_i}(b_i) = M_{\sigma_i}(P)(b_i) = 1$. Thus, $b \models R(\mathbf{G}^f, true)$ as there exists a path in $\mathbf{G}^f$ starting from an initial basic state tree $b_o \models P_o^f$ to b.

Therefore, $R(\mathbf{G}^f, true) = R(\mathbf{G}, P)$ as required.

Remarks

1. From the above proof, we have that the controlled system $\mathbf{G}^f$ is decided by P as follows.
   a) $P_o^f = P \wedge P_o$, and
   b) $(\forall \sigma \in \Sigma_c) f_\sigma := M_\sigma(P)$.
2. Even though P may not be reachable, the weak controllability of P guarantees that its reachable subpredicate $R(\mathbf{G}, P)$ can be implemented by a SFBC f. Thus the expensive computation of $R(\mathbf{G}, P)$ can be completely avoided.
3. The reverse statement in Theorem 3.1 does not hold. In Figure 3.2, the predicate P identified by the state set $\{0, 2\}$ is not weakly controllable, but there does exist a SFBC to implement $R(\mathbf{G}, P)$, identified by $\{0\}$.

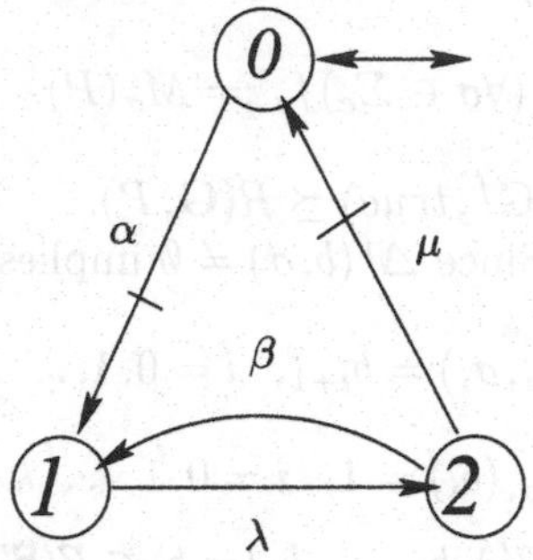

Fig. 3.2. A Counterexample

Now suppose P is not weakly controllable. Following the same line as in chapter 7 of [Won04], denote the family of weakly controllable subpredicates that are stronger than P by

$$\mathcal{CP}(P) := \{K \in \mathit{Pred}(\mathbf{ST}) | K \preceq P \;\&\; K \text{ weakly controllable } \}.$$

Proposition 3.2 $\mathcal{CP}(P)$ is nonempty and is closed under arbitrary disjunctions. In particular $\mathcal{CP}(P)$ contains a (unique) supremal element $\sup \mathcal{CP}(P)$.

Proof. We have that $\mathcal{CP}(P)$ is nonempty as $\perp \in \mathcal{CP}(P)$. Now let $K_\lambda \in \mathcal{CP}(P)$ for $\lambda \in \Lambda$, some index set. We need to show that

$$K := \bigvee_{\lambda \in \Lambda} K_\lambda \quad \in \mathcal{CP}(P).$$

It is clear that $K \preceq P$. So it remains to prove that K is weakly controllable. Let $b \models K$, i.e., $b \models K_\lambda$, for some λ. From the weak controllability of K_λ, $\forall \sigma \in \Sigma_u$,

$$b \models M_\sigma(K_\lambda) \preceq M_\sigma(K)$$

as $M_\sigma(\cdot)$ is monotone. Thus K is weakly controllable as required. Finally, the supremal element of $\mathcal{CP}(P)$ is given by

$$\sup \mathcal{CP}(P) = \bigvee \{K | K \in \mathcal{CP}(P)\}.$$

To compute $\sup \mathcal{CP}(P)$, we need to define a predicate transformer $[\cdot]$ in **G** by $R \mapsto [R]$, according to the inductive definition:

1. $b \models R \Rightarrow b \models [R]$;
2. $b \models [R]$ & $b \neq \emptyset$ & $\sigma \in \Sigma_u$ & $\Delta(b', \sigma) = b \Rightarrow b' \models [R]$;
3. No other basic state trees b satisfy $[R]$.

In effect, $[R]$ holds all the basic state trees that can reach R only by uncontrollable paths. We illustrate $[\cdot]$ in Figure 3.3. Evidently, $R \preceq [R]$.

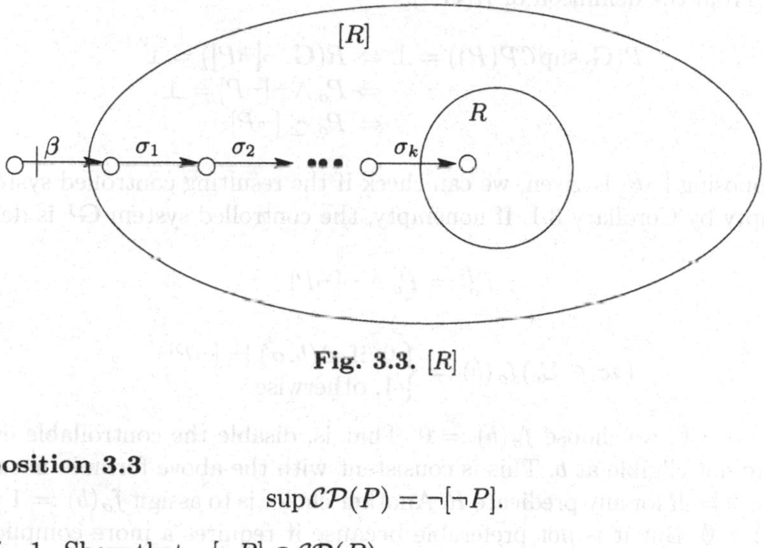

Fig. 3.3. $[R]$

Proposition 3.3

$$\sup \mathcal{CP}(P) = \neg[\neg P].$$

Proof. 1. Show that $\neg[\neg P] \in \mathcal{CP}(P)$.
We need to prove two claims.
a) $\neg[\neg P] \preceq P$.
This follows directly from the fact that $\neg P \preceq [\neg P]$.
b) $\neg[\neg P]$ is weakly controllable.
Let $b \models \neg[\neg P]$. Let $\sigma \in \Sigma_u$ and $b' = \Delta(b, \sigma) \neq \emptyset$. Suppose $b' \models [\neg P]$. Then by the definition of $[\cdot]$, $b \models [\neg P]$ as well, and we have a contradiction. So $b' \models \neg[\neg P]$, i.e., $b \models M_\sigma(\neg[\neg P])$.

2. Show that $P' \in \mathcal{CP}(P) \Rightarrow P' \preceq \neg[\neg P]$.
We need to show $b \models P' \Rightarrow b \models \neg[\neg P]$. Suppose $b \models [\neg P]$. Then for some $k \geq 0$, there exist $b_o (= b), b_1, \ldots, b_k$ and $\sigma_o, \sigma_1, \ldots, \sigma_{k-1} \in \Sigma_u$ such that

$$\Delta(b_i, \sigma_i) = b_{i+1} \neq \emptyset, \quad i = 0, 1, \ldots, k-1$$

and

$$b_k \models \neg P. \tag{3.1}$$

Also, P' is weakly controllable. Thus $b_o = b \models P'$ implies $b_1 \models P'$, for σ_o is uncontrollable and $b_o \models M_{\sigma_o}(P')$. Repetition yields

$$b_k \models P' \preceq P. \tag{3.2}$$

But (3.2) contradicts (3.1). Thus $b \models \neg[\neg P]$ as required.

We have the following corollary to verify if there exists a non-false controller.

Corollary 3.1

$$R(\mathbf{G}, \sup\mathcal{CP}(P)) = \bot \text{ iff } P_o \preceq [\neg P].$$

Proof. From the definition of $R(\mathbf{G}, \cdot)$,

$$\begin{aligned} R(\mathbf{G}, \sup\mathcal{CP}(P)) = \bot &\Leftrightarrow R(\mathbf{G}, \neg[\neg P]) = \bot \\ &\Leftrightarrow P_o \wedge \neg[\neg P] = \bot \\ &\Leftrightarrow P_o \preceq [\neg P]. \end{aligned}$$

Supposing $[\neg P]$ is given, we can check if the resulting controlled system is nonempty by Corollary 3.1. If nonempty, the controlled system $\mathbf{G}^f$ is defined by

$$P_o^f := P_o \wedge \neg[\neg P]$$

and

$$(\forall \sigma \in \Sigma_c) f_\sigma(b) := \begin{cases} 0, \text{ if } \Delta(b, \sigma) \models [\neg P] \\ 1, \text{ otherwise} \end{cases}.$$

If $\Delta(b, \sigma) = \emptyset$, we choose $f_\sigma(b) := 0$. That is, disable the controllable events that are not eligible at b. This is consistent with the above formula naturally because $\emptyset \models R$ for any predicate R. Another choice is to assign $f_\sigma(b) := 1$ when $\Delta(b, \sigma) = \emptyset$. But it is not preferable because it requires a more complicated formula for f_σ.

The following property of $[\cdot]$ is crucial to the parallel computing of $\sup\mathcal{CP}(P)$.

Proposition 3.4 $[P_1 \vee P_2] = [P_1] \vee [P_2]$, namely the following diagram commutes

Proof. 1. Show that $[P_1] \vee [P_2] \preceq [P_1 \vee P_2]$.
This follows directly from the fact that $P_i \preceq P_1 \vee P_2, i = 1, 2$ and $[\cdot]$ is monotone.

2. Show that $[P_1 \vee P_2] \preceq [P_1] \vee [P_2]$.
Let $b \models [P_1 \vee P_2]$. That is, for some $k \geq 0$, there exist basic state trees $b_o(= b), b_1, \ldots, b_k$ and $\sigma_o, \sigma_1, \ldots, \sigma_{k-1}$ such that

$$\Delta(b_i, \sigma_i) = b_{i+1}, \quad i = 0, 1, \ldots, k-1$$

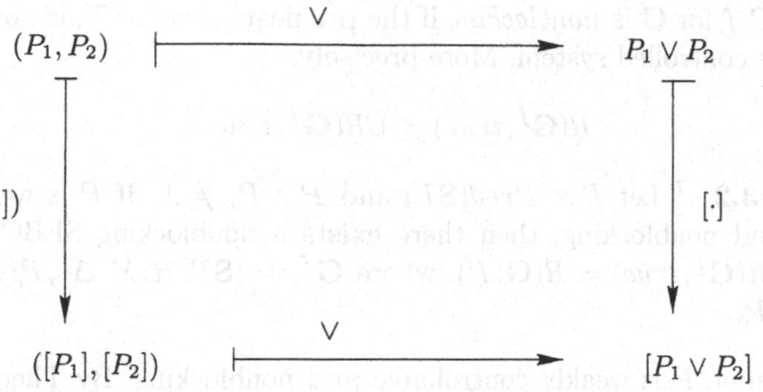

and

$$b_k \models P_1 \vee P_2.$$

This is the same as to say

$$b_k \models P_1 \text{ or } b_k \models P_2.$$

Thus, $b \models [P_1] \vee [P_2]$ as required.

The proof was easy. Furthermore, since the predicate $\neg P$ can usually be expressed as the disjunction of a set of simpler predicates R_i, for i in some index set, the above proposition states the important fact that the synthesis of each simpler R_i can be done independently. That is, a natural parallel algorithm is available.

3.3 Coreachability and Nonblocking Control

Let $\mathbf{G} = (\mathbf{ST}, \mathcal{H}, \Sigma, \Delta, P_o, P_m)$ be a state tree structure. Let $P \in Pred(\mathbf{ST})$. The *coreachability predicate* $CR(\mathbf{G}, P)$ is defined to hold for all basic state trees that can reach some $b_m \models P_m$ via trees satisfying P, according to the inductive definition:

1. $P_m \wedge P = false \Rightarrow CR(\mathbf{G}, P) = false$;
2. $(b_m \models P_m \wedge P) \Rightarrow (b_m \models CR(\mathbf{G}, P))$;
3. $b \models CR(\mathbf{G}, P) \;\&\; \sigma \in \Sigma \;\&\; \Delta(b', \sigma) = b \;\&\; b' \models P \Rightarrow b' \models CR(\mathbf{G}, P)$;
4. No other basic state trees satisfy $CR(\mathbf{G}, P)$.

For fixed $\mathbf{G}$, $CR(\mathbf{G}, \cdot) : Pred(\mathbf{ST}) \to Pred(\mathbf{ST})$ is a predicate transformer. Clearly $CR(\mathbf{G}, P) \preceq P$ and $CR(\mathbf{G}, \cdot)$ is monotone.

A predicate P is *nonblocking* for $\mathbf{G}$ if

$$R(\mathbf{G}, P) \preceq CR(\mathbf{G}, P),$$

i.e., each basic state tree, reachable from an initial basic state tree by a path on which all basic state trees satisfy P, can reach a marker basic state tree by a path on which all trees satisfy P.

A SFBC f for $\mathbf{G}$ is *nonblocking* if the predicate *true*, or $\top$, is nonblocking for $\mathbf{G}^f$, the controlled system. More precisely,

$$R(\mathbf{G}^f, true) \preceq CR(\mathbf{G}^f, true).$$

Theorem 3.2 [7] Let $P \in Pred(\mathbf{ST})$ and $P \wedge P_o \not\equiv \bot$. If P is weakly controllable and nonblocking, then there exists a nonblocking SFBC f for $\mathbf{G}$ such that $R(\mathbf{G}^f, true) = R(\mathbf{G}, P)$, where $\mathbf{G}^f := (\mathbf{ST}, \mathcal{H}, \Sigma, \Delta^f, P_o^f, P_m)$ and $P_o^f = P \wedge P_o$.

Proof. Assume P is weakly controllable and nonblocking. By Theorem 3.1, there exists a SFBC f such that $R(\mathbf{G}^f, true) = R(\mathbf{G}, P)$ if we define f by

$$(\forall \sigma \in \Sigma_c) f_\sigma := M_\sigma(P).$$

It remains to show that the same f is nonblocking as well, i.e.,

$$R(\mathbf{G}^f, true) \preceq CR(\mathbf{G}^f, true).$$

Let $b \models R(\mathbf{G}^f, true)$. Notice that $R(\mathbf{G}, P) = R(\mathbf{G}^f, true)$. So $b \models R(\mathbf{G}, P)$. Since P is nonblocking, $b \models CR(\mathbf{G}, P)$. That is, for some $k \geq 0$, there exist basic state trees $b_o(= b), b_1, \ldots, b_k$ and $\sigma_o, \sigma_1, \ldots, \sigma_{k-1} \in \Sigma$ such that

$$\Delta(b_i, \sigma_i) = b_{i+1}, \quad i = 0, 1, \ldots, k-1,$$

$$b_i \models P, \quad i = 0, 1, \ldots, k,$$

$$b_k \models P_m,$$

which implies $b_i \models M_{\sigma_i}(P)$, $i = 0, 1, \ldots, k-1$. Thus, we have $\Delta^f(b_i, \sigma_i) = b_{i+1}$, $i = 0, 1, \ldots, k-1$ as well, for if $\sigma_i \in \Sigma_u$, $f_{\sigma_i}(b_i) = 1$ since f is SFBC; while if $\sigma_i \in \Sigma_c$, $f_{\sigma_i}(b_i) = M_{\sigma_i}(P)(b_i) = 1$. Therefore $b \models CR(\mathbf{G}^f, true)$ as required.

Remarks

1. From the proof, we have that the given SFBC f guarantees

$$CR(\mathbf{G}, P) \preceq CR(\mathbf{G}^f, true),$$

   but not the other way around. This means in general, that there may exist redundant basic state trees in the controlled system $\mathbf{G}^f$ which do not satisfy $CR(\mathbf{G}, P)$. But we don't care about those basic state trees because they are not reachable (because P and f are nonblocking).

[7] A similar result is already given in chapter 7 of [Won04]. What is new in this theorem is again from the introduction of *weakly controllable* predicate.

A predicate P is *coreachable* in $\mathbf{G}$ if

$$P \preceq CR(\mathbf{G}, P).$$

Notice that the coreachability of P implies the nonblocking of P, as $R(\mathbf{G}, P) \preceq P$ in general. Also verifying the coreachability of P requires less computation, as we do not need to compute $R(\mathbf{G}, P)$. That is why we introduce Theorem 3.3.

Theorem 3.3 Let $P \in Pred(\mathbf{ST})$ and $P \wedge P_o \neq \bot$. Then there exists a nonblocking SFBC f for $\mathbf{G}$ such that $R(\mathbf{G}^f, true) = R(\mathbf{G}, P)$ if P is weakly controllable and coreachable, where $\mathbf{G}^f := (\mathbf{ST}, \mathcal{H}, \Sigma, \Delta^f, P_o^f, P_m)$ and $P_o^f = P \wedge P_o$.

Proof. Directly from the fact that coreachability implies nonblocking and Theorem 3.2.

Again suppose P is not weakly controllable or not coreachable. Denote the family of weakly controllable and coreachable subpredicates that are stronger than P by

$$\mathcal{C}^2\mathcal{P}(P) := \{K \in Pred(\mathbf{ST}) | K \preceq P \,\&\, K \text{ weakly controllable} \,\&\, K \text{ coreachable}\}$$

Proposition 3.5 $\mathcal{C}^2\mathcal{P}(P)$ is nonempty and is closed under arbitrary disjunctions. In particular $\mathcal{C}^2\mathcal{P}(P)$ contains a (unique) supremal element $\sup \mathcal{C}^2\mathcal{P}(P)$.

Proof. We have $\mathcal{C}^2\mathcal{P}(P)$ is nonempty as $\bot \in \mathcal{C}^2\mathcal{P}(P)$. Now let $K_\lambda \in \mathcal{C}^2\mathcal{P}(P)$ for $\lambda \in \Lambda$, some index set. We need to show that

$$K := \bigvee_{\lambda \in \Lambda} K_\lambda \quad \in \mathcal{C}^2\mathcal{P}(P).$$

Following the same line as in the proof for Proposition 3.2, we know $K \preceq P$ and K is weakly controllable. Similarly, let $b \models K$. So $b \models K_\lambda$, for some λ. By coreachability of K_λ, $b \models CR(\mathbf{G}, K_\lambda)$, and as $K_\lambda \preceq K$, we have $b \models CR(\mathbf{G}, K)$. Thus K is coreachable.

Finally, the supremal element of $\mathcal{C}^2\mathcal{P}(P)$ is given by

$$\sup \mathcal{C}^2\mathcal{P}(P) = \bigvee \{K | K \in \mathcal{C}^2\mathcal{P}(P)\}.$$

To compute $\sup \mathcal{C}^2\mathcal{P}(P)$, define a predicate transformer $\Omega_P : Pred(\mathbf{ST}) \to Pred(\mathbf{ST})$ in $\mathbf{G}$ by

$$\Omega_P(K) := P \wedge CR(\mathbf{G}, \sup \mathcal{CP}(K)) = P \wedge CR(\mathbf{G}, \neg[\neg K]).$$

Clearly, $\Omega_P(K) \preceq K$ and $\Omega_P(\cdot)$ is monotone, i.e., $K_1 \preceq K_2 \Rightarrow \Omega_P(K_1) \preceq \Omega_P(K_2)$.

Proposition 3.6 Let $S = \sup \mathcal{C}^2\mathcal{P}(P)$. Then

1. $S = \Omega_P(S)$, and
2. $(\forall K) K = \Omega_P(K) \Rightarrow K \preceq S$.

That is, S is the largest fixpoint of $\Omega_P(\cdot)$.

Proof. 1. Show that $S = \Omega_P(S)$.
$S \in \mathcal{C}^2\mathcal{P}(P)$ implies (i) $S \preceq P$; (ii) S is weakly controllable, i.e., $S = \sup \mathcal{CP}(S)$; and (iii) S is coreachable, i.e., $S = CR(G, S)$. Thus,

$$\begin{aligned}
\Omega_P(S) &:= P \wedge CR(\mathbf{G}, \sup \mathcal{CP}(S)) \\
&= P \wedge CR(\mathbf{G}, S), && \text{by (ii)} \\
&= P \wedge S, && \text{by (iii)} \\
&= S, && \text{by (i)}
\end{aligned}$$

2. Show that $(\forall K) K = \Omega_P(K) \Rightarrow K \preceq S$.
By definition, $K = \Omega_P(K) = P \wedge CR(\mathbf{G}, \sup \mathcal{CP}(K))$. Thus,
  a) $K \preceq P$;
  b) $K \preceq CR(\mathbf{G}, \sup \mathcal{CP}(K)) \preceq CR(\mathbf{G}, K)$, i.e., K is coreachable;
  c) $K \preceq CR(\mathbf{G}, \sup \mathcal{CP}(K)) \preceq \sup \mathcal{CP}(K) \Rightarrow K = \sup \mathcal{CP}(K)$, i.e., K is weakly controllable.
Therefore $K \in \mathcal{C}^2\mathcal{P}(P)$ and $K \preceq \sup \mathcal{C}^2\mathcal{P}(P) = S$.

Now we can compute $\sup \mathcal{C}^2\mathcal{P}(P)$ by iteration. Let

1. $K_o := P$, [8] and
2. $K_{i+1} = \Omega_P(K_i)$.

Then we have

Proposition 3.7 $K := \lim_{i \to \infty} K_i$ exists and $\sup \mathcal{C}^2\mathcal{P}(P) = K$.

Proof. Notice that $K_1 = \Omega_P(K_o) \preceq K_o$. Repeatedly, we have

$$\ldots K_j \preceq K_{j-1} \preceq \ldots K_1 \preceq K_o.$$

As *Pred*(**ST**) has a finite number of elements, there exists a finite number m, such that

$$\forall j < m, \;\; K_{j+1} \prec K_j$$

and

$$\forall j \geq m, \;\; K_j = K_m$$

Thus

$$K := \lim_{i \to \infty} K_i = K_m$$

exists and is the largest fixpoint of $\Omega_P(\cdot)$. Then by Proposition 3.6, $\sup \mathcal{C}^2\mathcal{P}(P) = K$.

Now we can give the most important algorithm in this chapter.

[8] We do not need to start with $K_o := true$ as we know $\sup \mathcal{C}^2\mathcal{P}(P) \preceq P$.

Algorithm 3.1

1. Let $K_o := P$.
2. $K_{i+1} := \Omega_P(K_i)$.
3. If $K_{i+1} = K_i$, then $\sup \mathcal{C}^2\mathcal{P}(P) = K_i$. Otherwise go back to step 2.

◇

The above algorithm will terminate because of Proposition 3.7. Each computation of $\Omega_P(K_i)$ needs to call $[\cdot]$ and $CR(\mathbf{G}, \cdot)$ once. So the faster these two functions execute, the better. We will come back to them later.

After having $\sup \mathcal{C}^2\mathcal{P}(P)$, the control is given by

$$P_o^f = P_o \wedge \sup \mathcal{C}^2\mathcal{P}(P)$$

and if $P_o^f \neq \perp$,

$$(\forall \sigma \in \Sigma_c) f_\sigma(b) := \begin{cases} 0, \text{ if } \Delta(b,\sigma) \models \neg \sup \mathcal{C}^2\mathcal{P}(P) \\ 1, \text{ otherwise} \end{cases}.$$

If $\Delta(b, \sigma) = \emptyset$, we choose $f_\sigma(b) := 0$. That is, disable the controllable events that are not eligible at b. This is naturally consistent with the above formula because $\emptyset \models R$ for any predicate R. Another choice is to assign $f_\sigma(b) := 1$ when $\Delta(b, \sigma) = \emptyset$. But it is not preferable because it requires a more complicated formula for f_σ.

So far, so good. But we still need to answer an important question. Notice that our $\sup \mathcal{C}^2\mathcal{P}(P)$ is the optimal supervisor that is coreachable instead of nonblocking. What if we define

$$\mathcal{NBCP}(P) := \{K \in Pred(\mathbf{ST}) | K \preceq P \;\&\; K \text{ weakly controllable } \&\; K \text{ nonblocking}\}.$$

Then following a similar routine, we can also prove the existence of $\sup \mathcal{NBCP}(P)$. So the question is:

Will $\sup \mathcal{C}^2\mathcal{P}(P)$ have the *same* control effect over $\mathbf{G}$ as $\sup \mathcal{NBCP}(P)$?

Or more precisely:

$$\text{is } R(\mathbf{G}, \sup \mathcal{C}^2\mathcal{P}(P)) = R(\mathbf{G}, \sup \mathcal{NBCP}(P))?$$

Proposition 3.8

$$R(\mathbf{G}, \sup \mathcal{C}^2\mathcal{P}(P)) = R(\mathbf{G}, \sup \mathcal{NBCP}(P)).$$

Proof. Write $S = \sup \mathcal{C}^2\mathcal{P}(P)$ and $T = \sup \mathcal{NBCP}(P)$.

1. Show that $R(\mathbf{G}, S) \preceq R(\mathbf{G}, T)$.
   S is coreachable, therefore nonblocking. So $S \in \mathcal{NBCP}(P)$ and $S \preceq T$. Thus $R(\mathbf{G}, S) \preceq R(\mathbf{G}, T)$ as $R(\mathbf{G}, \cdot)$ is monotone.

2. Show that $R(\mathbf{G},T) \preceq R(\mathbf{G},S)$.
   Let $b \models R(\mathbf{G},T)$. Then for some $k \geq 0$, there exist basic state trees $b_o, b_1, \ldots, b_k(=b)$ and $\sigma_o, \sigma_1, \ldots, \sigma_{k-1} \in \Sigma$ such that

$$b_o \models P_o; b_i \models T, \; i = 0,1,\ldots,k;$$

$$\Delta(b_i,\sigma_i) = b_{i+1}, \; i = 0,1,\ldots,k-1.$$

   As T is nonblocking and $b \models R(\mathbf{G},T)$, $b \models CR(\mathbf{G},T)$, i.e, there exist basic state trees $b_{k+1}, b_{k+2}, \ldots, b_m$ and $\sigma_k, \sigma_{k+1}, \ldots, \sigma_{m-1} \in \Sigma$ such that

$$b_m \models P_m; b_j \models T, \; j = k, k+1, \ldots, m;$$

$$\Delta(b_j,\sigma_j) = b_{j+1}, \; j = k, k+1, \ldots, m-1.$$

   Notice that $b \models R(\mathbf{G},T)$, $\Delta(b,\sigma_k) = b_{k+1}$ and $b_{k+1} \models T$. Thus $b_{k+1} \models R(\mathbf{G},T)$. By repeated application, we get

$$b_j \models R(\mathbf{G},T), \; j = k,\ldots,m.$$

   So $b_m \models P_m$, and thus $b \models CR(\mathbf{G}, R(\mathbf{G},T))$, i.e.,

$$R(\mathbf{G},T) \preceq CR(\mathbf{G}, R(\mathbf{G},T)),$$

   namely $R(\mathbf{G},T)$ is coreachable. Also $R(\mathbf{G},T) \preceq P$ and $R(\mathbf{G},T)$ is controllable (by Proposition 3.1). So $R(\mathbf{G},T) \in \mathcal{C}^2\mathcal{P}(P)$ and therefore $R(\mathbf{G},T) \preceq S = \sup \mathcal{C}^2\mathcal{P}(P)$. Apply $R(\mathbf{G},\cdot)$ on both sides, we have

$$R(\mathbf{G},T) = R(\mathbf{G}, R(\mathbf{G},T)) \preceq R(\mathbf{G},S)$$

   as required.

Notice that $\sup \mathcal{NBCP}(P)$ is by definition an optimal nonblocking controller. Proposition 3.8 states that the system $\mathbf{G}$ under the supervision of $\sup \mathcal{C}^2\mathcal{P}(P)$ has the same controlled behavior as that of $\sup \mathcal{NBCP}(P)$. Both controllers agree on their reachable subpredicates, even though they may differ on their nonreachable subpredicates. Therefore, $\sup \mathcal{C}^2\mathcal{P}(P)$ is another optimal nonblocking controller. The advantage of $\sup \mathcal{C}^2\mathcal{P}(P)$ is that there is no need to compute $R(\mathbf{G},\cdot)$.

3.4 Summary

The results in this chapter can be applied to the original RW framework. The contribution in our setting is that the onerous requirement of reachability in [Won04] can now be dropped, with no change in the resulting controlled behavior, but with a gain in computational efficiency for large systems.

Computing the coreachability subpredicate $CR(\mathbf{G},P)$ of P is as expensive as computing its reachability subpredicate. We can not avoid it in nonblocking control design. In the next chapter, we will spend more time on how to apply structural information to make it more efficient.

A predicate is descriptively no different from a set (in extensional form) if the system model does not have any particular structure. In that case, there is no better way to represent a predicate P than by enumerating all of the elements satisfying P. However, the results in this chapter will enjoy elegant application to our STS model because of its richness in structure.

4

Symbolic Computation of State Tree Structures

In the previous chapter, we developed a new algorithm for the nonblocking supervisory control of STS. The efficiency of this algorithm depends on the fast computation of $[\cdot]$ and $CR(\mathbf{G}, \cdot)$. In this chapter, we will exploit the rich structure of our STS models and symbolically compute $[\cdot]$ and $CR(\mathbf{G}, \cdot)$.

A sub-state-tree is a symbolic representation of its basic state tree set, which is equivalent to the flat state set. Because of the limited expressive power of state trees, however, there is in general no state tree representation of the union of two basic state tree sets. To work around this problem, we introduce symbolic computation using predicates directly. The predicate will be represented and manipulated by a BDD (Binary Decision Diagram) package called *BuDDy 2.0*. An alternative could be an IDD (Integer Decision Diagram) package if it could effectively handle the operators appearing in this chapter, e.g., the universal operator $(\forall v_x)P$ and the existential operator $(\exists v_x)P$.

Symbolic logic is one of the best ways to represent a transition system. In this chapter, we will work on the details of the symbolic STS model and symbolic synthesis. First we introduce the control problem for STS in section 4.1. In section 4.2, we give one symbolic representation of STS, then introduce the symbolic synthesis in section 4.3, and discuss the implementation of synthesis and controller in section 4.4. Finally, we give two tutorial examples in section 4.5 and a summary in section 4.6 to close the chapter.

4.1 Control Problem for STS

Let $\mathbf{G} = (\mathbf{ST}, \mathcal{H}, \Sigma, \Delta, \mathbf{ST}_o, \mathcal{ST}_m)$ be a state tree structure with $\mathbf{ST} = (X, x_o, \mathcal{T}, \mathcal{E})$. A specification $\mathcal{S}$ is simply given as a set of *illegal* sub-state-trees, which the supervisor must forbid the system to visit. Formally,

$$\mathcal{S} \subseteq \mathcal{ST}(\mathbf{ST}).$$

Let the predicate P be the characteristic function of the set of basic state trees $\bigcup_{S \in \mathcal{S}} \mathcal{B}(S)$. Then the *control problem* is to find the supremal element

$$\sup \mathcal{C}^2\mathcal{P}(\neg P).$$

We call the pair $(\mathbf{G}, \mathcal{S})$, or $(\mathbf{G}, P)$, a *control problem.* A special case is when $\mathcal{S}(=\emptyset)$ is empty. Then the control problem $(\mathbf{G}, \emptyset)$, or $(\mathbf{G}, false)$, is just to make the system nonblocking.

Figure 4.1 illustrates how to write such a specification. The two machines, M_1 and M_2, are not allowed to occupy the superstates W_1 and W_2 at the same time. The forbidden state specification is given in (b) of Figure 4.1. We can write the specification state tree in (b) by the set of its active states (refer to page 27 for *active*)

$$\{W_1, W_2\},$$

which is a simpler form. In fact, the predicate version of this specification is also simple. Using the Θ function defined in the next section, one can use the predicate $P := (v_{M_1} = W_1) \wedge (v_{M_2} = W_2)$, where v_{M_1} and v_{M_2} are the state variables for M_1 and M_2, respectively. This specification is *explicitly* saying what the end users asked for, i.e., the dangerous situation (to be prohibited) is whenever M_1 and M_2 are at W_1 and W_2 at the same time. If one tries to write a generator specification for the same control problem, he has to build

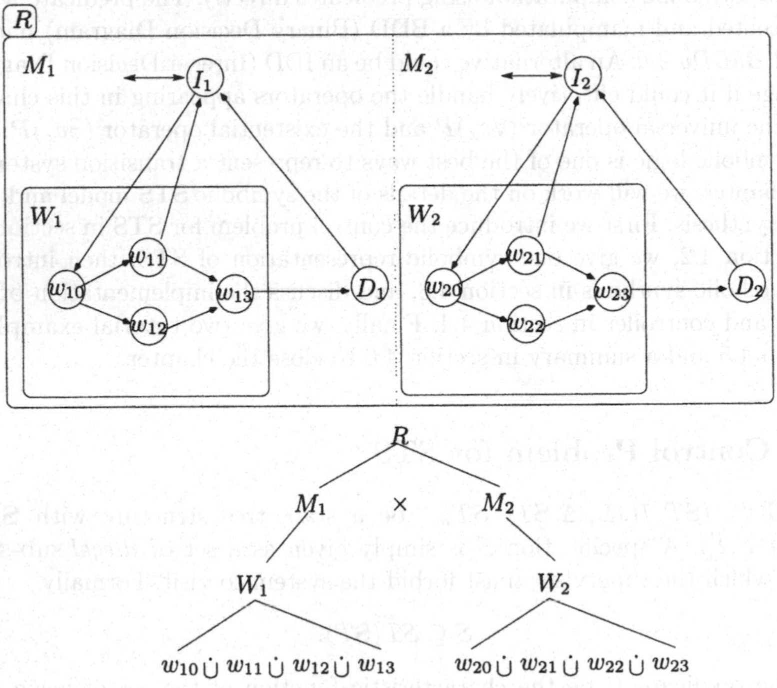

(b) specification state tree

Fig. 4.1. Forbidden state specification

a generator by synchronizing M_1 and M_2 to get the overall picture of this system and then delete those illegal states one by one. Even if one gets a generator specification eventually, it may be hard to convince the end users that it is *really* what they intend.

Our setting of control problems can also cover the so-called dynamic state feedback control (see chapter 7 of [Won04]), where memory modules are adjoined to the plant system. Let us first look at what we do in the original RW framework. A specification in the original RW framework is a generator, which acts as a memory to "remember" past behavior as well as a preliminary controller by synchronizing with the plant generator. The control requirement is given *implicitly* inside the generator. Briefly, a specification in the original RW framework includes

$$\text{memory} + \text{control}.$$

In our setting, we put the memory part into the STS model and try to write a predicate for the control part. Specifically, a *memory* or *specification holon* is just a holon $H = (X, \Sigma, \delta, X_o, X_m)$ in $\mathbf{G}$ such that $(\forall x \in X_I, \sigma \in \Sigma_u)\delta_I(x, \sigma)!$, i.e., it does not constrain any uncontrollable transitions. Memories or specification holons are introduced in $\mathbf{G}$ to record past behavior appearing in the control problem description. [1] So here $\mathbf{G}$ describes the necessary information required by the control problem, and therefore is accurately the STS model of *the control problem* instead of merely the uncontrolled plant.

Here we show how to treat the Small Factory with a buffer. First we write its generator specification in Figure 4.2. The specification BUF does two things. First it says that when β_1 occurs, a workpiece will be put in the

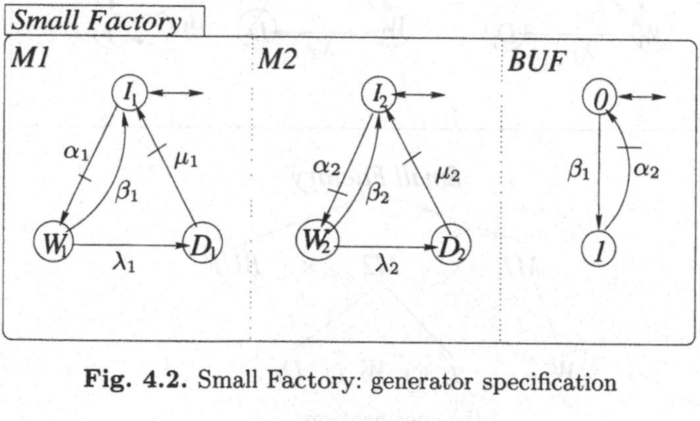

Fig. 4.2. Small Factory: generator specification

[1] $\mathbf{G}$ must be well-formed, which limits the types of memory we can have in $\mathbf{G}$. For example, the events of the inner transition of the memory can only be shared among the holons matched to the AND components that are AND-adjacent to the same AND superstate.

buffer and when α_2 occurs, the workpiece will be taken to machine $M2$. Then it implicitly describes the control part by

1. not allowing α_2 to occur at state 0 and
2. not allowing β_1 to occur at state 1.

In order to transform this problem into our setting, we need to find the *illegal* states implied in these two requirements. Notice that the second requirement violates controllability because β_1 is uncontrollable. So we can infer that the illegal state tree identified by the second requirement is given by the active state set

$$\{W_1, 1\}$$

because the only states that β_1 exits are W_1 and 1. In respect to α_2, one must be careful that no states will be rendered illegal from the first control requirement (disabling α_2 at state 0) because α_2 is controllable. The only effect of disabling α_2 is that the controlled system may now be blocking. In Figure 4.3, we show the STS model of the Small Factory, and its specification given as the illegal sub-state-tree in (b). This illegal sub-state-tree is due to the requirement of not allowing β_1 to occur at state 1. Because of the introduction of this illegal sub-state-tree as the specification, we can now add the selfloop of β_1 at state 1 in BUF. That is, BUF can be modelled as a memory.

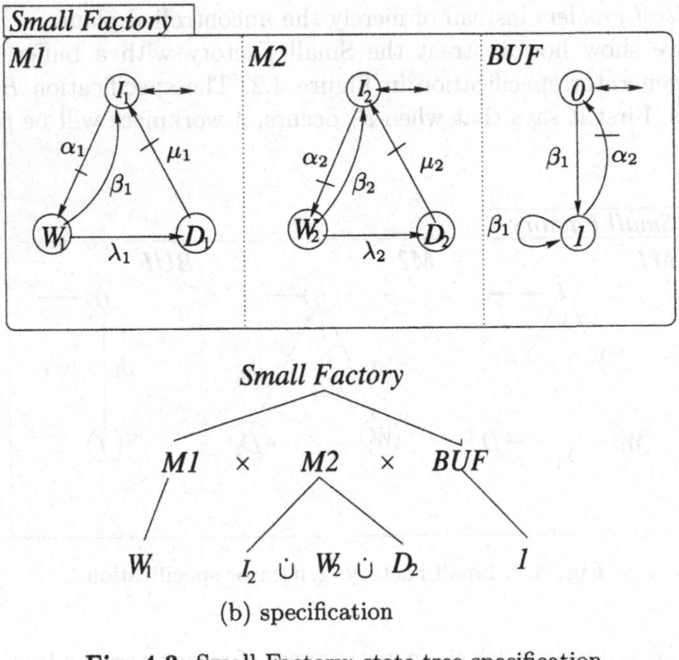

(b) specification

Fig. 4.3. Small Factory: state tree specification

The specification state tree shown in (b) is simple. But it is still not the *same* as what the end users asked for, which is not allowing β_1 to occur at state 1. So we introduce another type of specification for this requirement. Let

$$\mathcal{S}' := \{(S, \sigma) | S \in \mathcal{ST}(\mathbf{ST}) \; \& \; \sigma \in \Sigma_u\}.$$

We can compute the illegal sub-state-tree from each pair in $\mathcal{S}'$. For example, in the above Small Factory, the specification can be written as

$$\mathcal{S}' := \{(S, \beta_1)\},$$

where S is given by the active state set $\{1\}$. This is exactly what the end users asked for: preventing β_1 at state 1 of the holon BUF. The resulting illegal state tree $\{W_1, 1\}$ illustrated in (b) of Figure 4.3 can be computed by

$$S \wedge \text{Elig}_{\textit{Small Factory}}(\beta_1),$$

where $\text{Elig}_{\textit{Small Factory}}(\beta_1)$, given by its active state set $\{W_1\}$, is the largest eligible state tree of β_1 on the STS *Small Factory* in Figure 4.3. [2] Figure 4.4 illustrates the above computation.

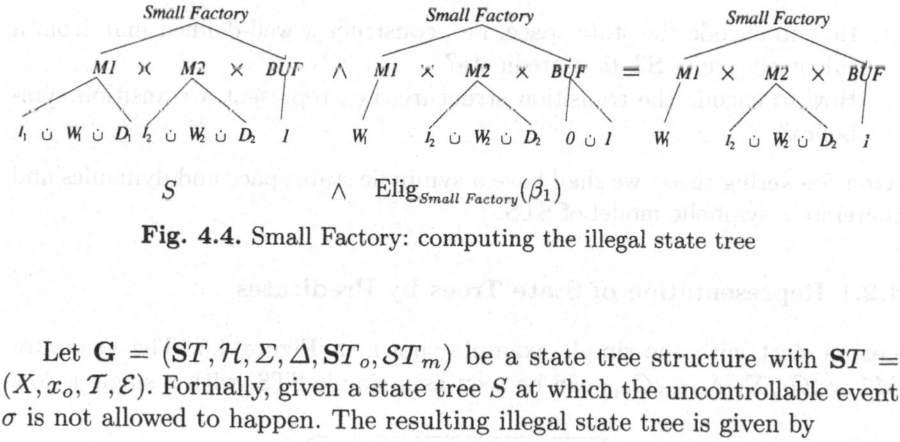

Fig. 4.4. Small Factory: computing the illegal state tree

Let $\mathbf{G} = (\mathbf{ST}, \mathcal{H}, \Sigma, \Delta, \mathbf{ST}_o, \mathcal{ST}_m)$ be a state tree structure with $\mathbf{ST} = (X, x_o, \mathcal{T}, \mathcal{E})$. Formally, given a state tree S at which the uncontrollable event σ is not allowed to happen. The resulting illegal state tree is given by

$$S \wedge \text{Elig}_{\mathbf{G}}(\sigma),$$

where $\text{Elig}_{\mathbf{G}}(\sigma)$ is the largest eligible state tree of σ on $\mathbf{G}$.

In summary, the control problem can be given as a triple $(\mathbf{G}, \mathcal{S}, \mathcal{S}')$ for the benefit of end users. From there we can simplify it to the original control problem $(\mathbf{G}, \mathcal{S})$ given at the beginning of this section by applying the above formula. Here our control problem is given in terms of *illegal* sub-state-trees. This is because the set of *illegal* sub-state-trees is usually much smaller than

---

[2] Recall subsection 2.3.1 on the Δ function in Chapter 2 for the definition of $\text{Elig}_{\mathbf{G}}(\sigma)$.

that of the *legal* sub-state-trees. One can also define a control problem by $(\mathbf{G}, R)$, where R is the predicate representing the set of *legal* sub-state-trees. Our synthesis approach is still applicable to this control problem. In this case, what we need is just to compute $\sup \mathcal{C}^2\mathcal{P}(R)$.

Sometimes we need to add memory modules to a given plant STS model, in order to describe some specifications. In the Small Factory, the BUF is a memory added to the plant STS model of two concurrent machines $M1$ and $M2$. In order for the extended STS model with memories still to be well-formed, we require that each memory respect the *local coupling* property. This is the price to pay in order to keep the structure. Otherwise one will have to "flatten" some levels to make the resulting STS well-formed.

Without losing generality, therefore, we discuss in the following sections only the synthesis of the original control problem $(\mathbf{G}, \mathcal{S})$.

4.2 Symbolic Representation of STS

Let $\mathbf{G} = (\mathbf{ST}, \mathcal{H}, \Sigma, \Delta, \mathbf{ST}_o, \mathcal{ST}_m)$ be a state tree structure with $\mathbf{ST} = (X, x_o, \mathcal{T}, \mathcal{E})$. There are two questions to be answered in this section.

1. How to encode the state space, i.e., construct a well-defined map from a sub-state-tree of $\mathbf{ST}$ to a predicate?
2. How to encode the transition structure, i.e., represent a transition symbolically?

After answering these, we shall have a symbolic state space and dynamics and therefore a symbolic model of STS.

4.2.1 Representation of State Trees by Predicates

Let us start with the simple example shown in Figure 4.5. The generator $M1 = (Q_1, \Sigma_1, \delta_1, q_o, Q_m)$ can be seen as a simple STS, with a single holon.

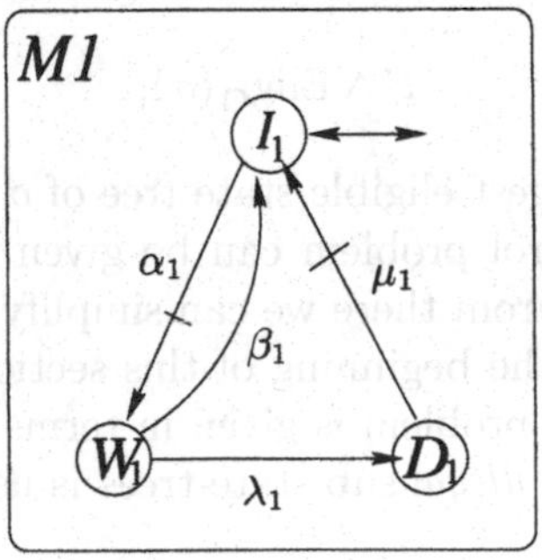

Fig. 4.5. Machine

$M1$ has 3 states, I_1(Idle), W_1(working) and D_1(down). Let $A \in 2^{Q_1}$, the power set of Q_1. Usually there are two ways to describe the set A. The first is just to enumerate all states inside A, say $A = \{I_1, W_1\}$. Another one is to define a predicate(or a *characteristic function*) $P_A : Q_1 \rightarrow \{0, 1\}$ over Q_1 such that $q \in A \Leftrightarrow P_A(q) = 1$. However, how to write such an explicit predicate (or function) that is accurate and understandable is a challenge.

Here we show one way to do it. First denote by v_{M1} the *state variable* of $M1$. That is, v_{M1} is a variable whose range is the internal state set Q_1 of holon $M1$. By introducing this state variable, we can write a logic formula as the predicate $P_A(v_{M1})$ of any given set A. One simple predicate here is $P_A(v_{M1}) := 1$. It is the predicate for the set $A = Q_1$ because, for any state $q \in Q_1$, $P_A(q) = 1$ from the definition. Another simple predicate is $P_A(v_{M1}) := 0$ for the empty set $A = \emptyset$ because for any state $q \in Q_1$, $P_A(q) = 0$. However, we cannot gain much if the set Q_1 itself does not have some regularity of mathematical structure. For example, if $A = \{I_1, W_1\}$, we can only write

$$P_A(v_{M1}) := (v_{M1} = I_1) \vee (v_{M1} = W_1).$$

In the formula, the operator $=$ in $(v_{M1} = I_1)$ returns value "1" if and only if v_{M1} has been assigned value I_1 (otherwise it returns value "0"). Because of this new meaning of $=$, from now on, we will use $\equiv$ for logical equivalence, i.e., when two predicates are equivalent. $P_A(v_{M1})$ is a valid predicate for the set $A = \{I_1, W_1\}$. To ask if $I_1 \in A$, we just need to assign I_1 to the state variable v_{M1} and evaluate the above formula as follows.

$$P_A(I_1) := (I_1 = I_1) \vee (I_1 = W_1) \equiv 1 \vee 0 \equiv 1.$$

It is enough to conclude $I_1 \in A$ because $P_A(I_1) \equiv 1$. However, this formula $P_A(v_{M1})$ is not any simpler than just enumerating all elements of A.

So symbolic representation of a flat model does not look promising. But it will be different if the model has structure. For example, the Small Factory in Figure 4.6 is the synchronous product of 3 generators, $M1$, $M2$ and BUF. It can also be looked on as a special STS with 3 AND components. Let Q_1, Q_2 and Q_3 be the state set of $M1, M2$ and BUF, respectively. Then we can define the characteristic function $P_A : Q_1 \times Q_2 \times Q_3 \rightarrow \{0, 1\}$ for each set $A \in 2^{Q_1 \times Q_2 \times Q_3}$, which is a function taking three arguments: v_{M1}, v_{M2} and v_{BUF}. The simplest ones are still $P_A(\mathbf{v}) := 1$ for the set $A = Q_1 \times Q_2 \times Q_3$ and $P_A(\mathbf{v}) := 0$ for the empty set $\emptyset$, where $\mathbf{v} = (v_{M1}, v_{M2}, v_{BUF})$ is the state variable set. However, now we may gain economy for other sets. For example, let

$$A = \left\{ \begin{array}{ll} (I_1, I_2, 0), & (I_1, I_2, 1), \\ (I_1, W_2, 0), & (I_1, W_2, 1), \\ (I_1, D_2, 0), & (I_1, D_2, 1) \end{array} \right\}.$$

Its characteristic function is as simple as

$$P_A(\mathbf{v}) := (v_{M1} = I_1).$$

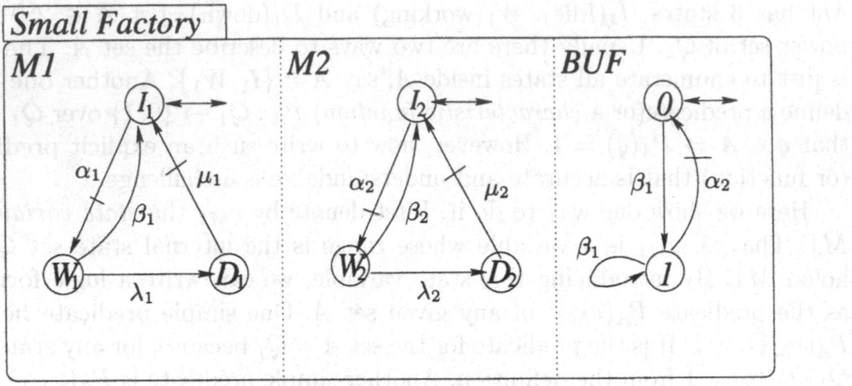

Fig. 4.6. Small Factory

For example, $(I_1, I_2, 0) \in A$ because

$$P_A(I_1, I_2, 0) := (I_1 = I_1) \equiv 1;$$

$(W_1, I_2, 0) \notin A$ because

$$P_A(W_1, I_2, 0) := (W_1 = I_1) \equiv 0.$$

The above demonstrates that using predicates has an advantage for systems modelled with structure. Now we come back to general STS models. Let $\mathbf{G} = (\mathbf{ST}, \mathcal{H}, \Sigma, \Delta, \mathbf{ST}_o, ST_m)$ be a state tree structure with $\mathbf{ST} = (X, x_o, \mathcal{T}, \mathcal{E})$. A sub-state-tree $\mathbf{ST}_1$ of $\mathbf{ST}$ is a symbolic representation of the basic state tree set $\mathcal{B}(\mathbf{ST}_1)$. So what we need to do is to build a unique predicate $P_{\mathbf{ST}_1}$ from $\mathbf{ST}_1$.

Definition 4.1 [State variable, Θ] A *state variable* for a holon H (or an OR superstate x) is a variable whose range is the internal state set of H (or the set of all children of x).

Assign a state variable to each OR superstate on the state tree $\mathbf{ST}$. Denote by v_x the state variable for the OR superstate x. Let $\mathbf{ST}_1 = (X_1, x_o, \mathcal{T}_1, \mathcal{E}_1)$ be a sub-state-tree of $\mathbf{ST}$. Define $\Theta : \mathcal{ST}(\mathbf{ST}) \longrightarrow \mathit{Pred}(\mathbf{ST})$ recursively by

$$\Theta(\mathbf{ST}_1) := \begin{cases} \bigwedge_{y \in \mathcal{E}_1(x_o)} \Theta'(\mathbf{ST}_1^y), & \text{if } \mathcal{T}(x_o) = \mathit{and} \\ \bigvee_{y \in \mathcal{E}_1(x_o)} ((v_{x_o} = y) \wedge \Theta'(\mathbf{ST}_1^y)), & \text{if } \mathcal{T}(x_o) = \mathit{or} \\ 1, & \text{if } \mathcal{T}(x_o) = \mathit{simple} \end{cases},$$

where $\mathbf{ST}_1^y$ denotes the child state tree of $\mathbf{ST}_1$ that is rooted by y, and assume that $\Theta' : \mathcal{ST}(\mathbf{ST}^y) \to \mathit{Pred}(\mathbf{ST}^y)$ is already defined on the child state tree $\mathbf{ST}^y$. By abusing notation, we can omit the $'$ and write

$$\Theta(\mathbf{ST}_1) := \begin{cases} \bigwedge_{y \in \mathcal{E}_1(x_o)} \Theta(\mathbf{ST}_1^y), & \text{if } \mathcal{T}(x_o) = \textit{and} \\ \bigvee_{y \in \mathcal{E}_1(x_o)} ((v_{x_o} = y) \wedge \Theta(\mathbf{ST}_1^y)), & \text{if } \mathcal{T}(x_o) = \textit{or} \\ 1, & \text{if } \mathcal{T}(x_o) = \textit{simple} \end{cases}$$

Call each term $(v_{x_o} = y)$ an *atomic term* of $\Theta(\mathbf{ST}_1)$. Trivially, define $\Theta(\mathbf{ST}_1) := 0$ if $\mathbf{ST}_1$ is the empty state tree.

Notice that we can exploit the following tautology [3]

$$(\bigvee_{y \in \mathcal{E}(x_o)} (v_{x_o} = y)) \equiv 1$$

to simplify $\Theta(\mathbf{ST}_1)$.

◇

Remarks

1. $\Theta(\mathbf{ST}_1)$ is a propositional logic formula. Let $\mathbf{v} := (v_1, v_2, \ldots, v_n)$ be the set of all state variables on $\mathbf{ST}$. Then $\Theta(\mathbf{ST}_1)$ can be considered an n-ary function, with $\mathbf{v}$ as its argument set. So we can also write $\Theta(\mathbf{ST}_1)(\mathbf{v})$ to emphasize the fact that it is a function.
2. Because of the introduction of $=$ in the atomic term of Θ, the logical equivalence of two predicates P and R will be written $P \equiv R$.
3. We demonstrate the computation of $\Theta(\mathbf{ST}_1)$ in Figure 4.7. For the sub-state-tree $\mathbf{ST}_1$ in (b), we have

$$\Theta(\mathbf{ST}_1) := \Theta(\mathbf{ST}_1^{y_1}) \wedge \Theta(\mathbf{ST}_1^{y_2}) \wedge \Theta(\mathbf{ST}_1^{y_3}).$$

Notice that

$$\begin{aligned} \Theta(\mathbf{ST}_1^{y_1}) &:= (v_{y_1} = z_1) \wedge \Theta(\mathbf{ST}_1^{z_1}) \vee (v_{y_1} = z_2) \wedge 1 \\ &\equiv (v_{y_1} = z_1) \wedge (v_{z_1} = a_1) \vee (v_{y_1} = z_2), \\ \Theta(\mathbf{ST}_1^{y_2}) &:= (v_{y_2} = z_3), \\ \Theta(\mathbf{ST}_1^{y_3}) &:= (v_{y_3} = z_5) \vee (v_{y_3} = z_6) \equiv 1. \end{aligned}$$

Thus

$$\Theta(\mathbf{ST}_1) \equiv [(v_{y_1} = z_1) \wedge (v_{z_1} = a_1) \vee (v_{y_1} = z_2)] \wedge (v_{y_2} = z_3).$$

Following the same routine, for the sub-state-tree $\mathbf{ST}_2$ in (c), we have

$$\begin{aligned} \Theta(\mathbf{ST}_2) &:= \Theta(\mathbf{ST}_2^{y_1}) \wedge \Theta(\mathbf{ST}_2^{y_2}) \wedge \Theta(\mathbf{ST}_2^{y_3}) \\ &\equiv 1 \wedge (v_{y_2} = z_3) \wedge 1 \\ &\equiv (v_{y_2} = z_3). \end{aligned}$$

[3] This says that the predicate $\Theta(\mathbf{ST}_1)$ is independent of the state variable v_{x_o} if all descendants of x_o are on the state tree. Recall that $\equiv$ means that the two formulas are logically equivalent. Notice that this tautology is *automatically* applied in building a BDD representation of a predicate, namely the BDD reduction rule will simplify the boolean formula $(v = 0 \vee v = 1)$ to the logical truth 1.

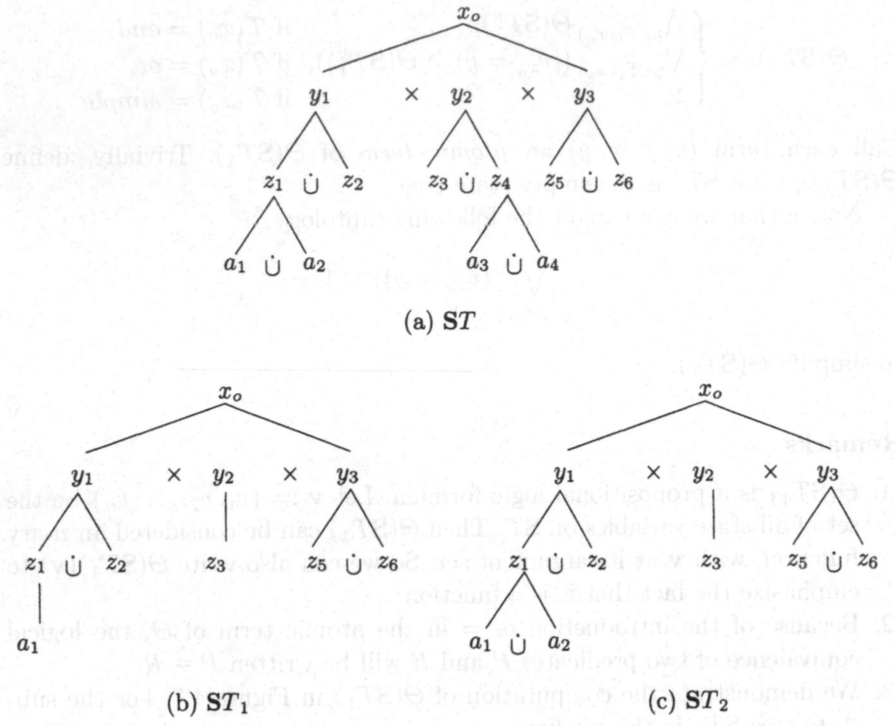

Fig. 4.7. Examples for Θ

4. From the above examples, we observe that the formula $\Theta(\mathbf{ST}_2)$ is simpler (shorter) because all descendants of y_1 are on $\mathbf{ST}_2$. This observation is the key reason that our structural algorithm will be depth first in order to control the size of predicates during the computation of $[\cdot]$ and $CR(\mathbf{G}, \cdot)$.
5. Also notice that $|\mathcal{B}(\mathbf{ST}_1)| < |\mathcal{B}(\mathbf{ST}_2)|$ $(4 < 6)$ even though the formula $\Theta(\mathbf{ST}_2)$ is shorter. This means that the length of the predicate formula is not closely correlated with the size of the set it represents. It is much more sensitive to the set's tree structure. A larger set may have a shorter logic formula as its characteristic function.

Let b be a basic sub-state-tree of **ST**. By writing $\Theta(\mathbf{ST}_1)(b)$ we mean: first assign the only child of each OR superstate x on b to its respective state variable v_x in $\Theta(\mathbf{ST}_1)$, [4] and if $\Theta(\mathbf{ST}_1)$ is a *valid predicate*, then the resulting logic formula should be simplified as either 1 or 0 (otherwise we cannot call $\Theta(\mathbf{ST}_1)$ a predicate). The following lemma guarantees that $\Theta(\mathbf{ST}_1)$ is a valid predicate.

Lemma 4.1 For a *basic* state tree b, it is always the case that either $\Theta(\mathbf{ST}_1)(b) \equiv 1$ or $\Theta(\mathbf{ST}_1)(b) \equiv 0$.

[4] On a *basic* state tree, every OR superstate has exactly one child. Otherwise its count will be more than 1.

Proof. The proof is from the definitions of both *basic* state tree and Θ. From the definition of Θ, we have

$$\Theta(\mathbf{ST}_1)(b) := \begin{cases} \bigwedge_{y\in\mathcal{E}_1(x_o)} \Theta(\mathbf{ST}_1^y)(b), & \text{if } \mathcal{T}(x_o) = \textit{and} \\ \bigvee_{y\in\mathcal{E}_1(x_o)} ((v_{x_o} = y) \wedge \Theta(\mathbf{ST}_1^y))(b), & \text{if } \mathcal{T}(x_o) = \textit{or} \\ 1, & \text{if } \mathcal{T}(x_o) = \textit{simple} \end{cases}$$

If $x_o \in b$ is an AND state, the evaluation of $\Theta(\mathbf{ST}_1)(b)$ is completely decided by all $\Theta(\mathbf{ST}_1^y)(b)$. If all of them are either 0 or 1, then $\Theta(\mathbf{ST}_1)(b)$ must be either 0 or 1.

If $x_o \in b$ is an OR state, then on the *basic* state tree b, x_o has a unique child, say y. If $y \notin \mathcal{E}_1(x_o)$, $\Theta(\mathbf{ST}_1)(b) \equiv 0$ from the definition, already decided in this level. Otherwise, if $y \in \mathcal{E}_1(x_o)$, $\Theta(\mathbf{ST}_1)(b) := \Theta(\mathbf{ST}_1^y))(b)$, again decided by the value on its child state tree.

So the value of $\Theta(\mathbf{ST}_1^y)(b)$ at any superstate y of b is either 0 or again decided by the return value on its child state trees. Finally, it must be decided because for any leaf state y on b, $\Theta(\mathbf{ST}_1^y)(b) := 1$ is decided.

The following lemma is also important.

Lemma 4.2 Let $\mathbf{G} = (\mathbf{ST}, \mathcal{H}, \Sigma, \Delta, \mathbf{ST}_o, \mathcal{ST}_m)$ be a state tree structure with $\mathbf{ST} = (X, x_o, \mathcal{T}, \mathcal{E})$. Let $\mathbf{ST}_1$ and $\mathbf{ST}_2$ be two sub-state-trees of $\mathbf{ST}$. We have

1. Θ is injective, i.e., $\Theta(\mathbf{ST}_1) \equiv \Theta(\mathbf{ST}_2) \Rightarrow \mathbf{ST}_1 = \mathbf{ST}_2$;
2. $\mathbf{ST}_1 \leq \mathbf{ST}_2 \Leftrightarrow \Theta(\mathbf{ST}_1) \preceq \Theta(\mathbf{ST}_2)$;
3. $\mathbf{ST}_1 \models \Theta(\mathbf{ST}_1)$.

Proof. 1. We show that $\Theta(\mathbf{ST}_1) \equiv \Theta(\mathbf{ST}_2) \Rightarrow \mathbf{ST}_1 = \mathbf{ST}_2$.
The prove is by contradiction. Assume $\mathbf{ST}_1 \neq \mathbf{ST}_2$. Then there must exist an OR component on only one of the two state trees, say $\mathbf{ST}_1$. Suppose there exists $y \in \mathcal{E}_1(x) - \mathcal{E}_2(x)$ for some OR superstate x. Then $\Theta(\mathbf{ST}_1^x) \not\equiv \Theta(\mathbf{ST}_2^x)$ because the atomic term $(v_x = y)$ only appears in $\Theta(\mathbf{ST}_1^x)$. Now for any ancestor z of x, including the root state x_o, we will have $\Theta(\mathbf{ST}_1^z) \not\equiv \Theta(\mathbf{ST}_2^z)$ because the state variable v_x for x only occurs inside $\Theta(\mathbf{ST}_1^x)$ and $\Theta(\mathbf{ST}_2^x)$. This contradicts $\Theta(\mathbf{ST}_1) \equiv \Theta(\mathbf{ST}_2)$.

2. We show that $\mathbf{ST}_1 \leq \mathbf{ST}_2 \Leftrightarrow \Theta(\mathbf{ST}_1) \preceq \Theta(\mathbf{ST}_2)$
   a) Show that $\mathbf{ST}_1 \leq \mathbf{ST}_2 \Rightarrow \Theta(\mathbf{ST}_1) \preceq \Theta(\mathbf{ST}_2)$ by recursion.

      i. (terminal case) Let y be a leaf state on $\mathbf{ST}_1$. Then y must also be a leaf state on $\mathbf{ST}_2$ because $\mathbf{ST}_1 \leq \mathbf{ST}_2$. Thus, $\Theta(\mathbf{ST}_1^y) \equiv 1 \preceq 1 \equiv \Theta(\mathbf{ST}_2^y)$.
      ii. (recursive cases) We need to show that $\Theta(\mathbf{ST}_1^{x_o}) \preceq \Theta(\mathbf{ST}_2^{x_o})$ if $(\forall y \in \mathcal{E}_1(x_o))\Theta(\mathbf{ST}_1^y) \preceq \Theta(\mathbf{ST}_2^y)$. This is obvious whenever $\mathcal{T}(x_o) = \textit{and}$ or $\mathcal{T}(x_o) = \textit{or}$, just because $\mathbf{ST}_1 \leq \mathbf{ST}_2 \Rightarrow \mathcal{E}_1(x_o) \subseteq \mathcal{E}_2(x_o)$.

   The proof will terminate as it will finally reach the terminal case.

b) Show that $\mathbf{ST}_1 \leq \mathbf{ST}_2 \Leftarrow \Theta(\mathbf{ST}_1) \preceq \Theta(\mathbf{ST}_2)$ by contradiction. Assume $\mathbf{ST}_1 \not\leq \mathbf{ST}_2$. Then there must exist at least one OR component y such that $y \in \mathcal{E}_1(x) - \mathcal{E}_2(x)$ for some OR superstate x. Then $\Theta(\mathbf{ST}_1^x) \not\preceq \Theta(\mathbf{ST}_2^x)$ because the atomic term $(v_x = y)$ only appears in $\Theta(\mathbf{ST}_1^x)$. Now for any ancestor z of x, including the root state x_o, we can have $\Theta(\mathbf{ST}_1^z) \not\preceq \Theta(\mathbf{ST}_2^z)$ because the state variable v_x for x only occurs inside $\Theta(\mathbf{ST}_1^x)$ and $\Theta(\mathbf{ST}_2^x)$. This contradicts $\Theta(\mathbf{ST}_1) \preceq \Theta(\mathbf{ST}_2)$.

3. Show that $\mathbf{ST}_1 \models \Theta(\mathbf{ST}_1)$.
   a) If $\mathbf{ST}_1$ is a basic state tree, then $\mathbf{ST}_1 \models \Theta(\mathbf{ST}_1)$ because all atomic terms in $\Theta(\mathbf{ST}_1)$ will give value 1 after assigning each OR component on $\mathbf{ST}_1$ to its respective state variable in $\Theta(\mathbf{ST}_1)$. Thus $\Theta(\mathbf{ST}_1)$ is the characteristic function of the set $\{\mathbf{ST}_1\}$ and $\Theta(\mathbf{ST}_1)(\mathbf{ST}_1) \equiv 1$ as required.
   b) If $\mathbf{ST}_1$ is not a basic state tree, then we need to prove $(\forall b \in \mathcal{B}(\mathbf{ST}_1)) b \models \Theta(\mathbf{ST}_1)$. Let $b \in \mathcal{B}(\mathbf{ST}_1)$. Then $b \models \Theta(b)$. Also $\Theta(b) \preceq \Theta(\mathbf{ST}_1)$ from the item 2 of this lemma. Therefore,

$$b \models \Theta(b) \preceq \Theta(\mathbf{ST}_1) \Rightarrow b \models \Theta(\mathbf{ST}_1)$$

   as required.

However, Θ is not surjective. This is because the expressiveness of state trees is limited and therefore in general, a predicate is equivalent to a set of sub-state-trees instead of just one sub-state-tree. One counterexample is the following. In the Small Factory of Figure 4.6 on page 84, given the predicate

$$P := (v_{M1} = I_1 \wedge v_{M2} = I_2) \vee (v_{M1} = D_1 \wedge v_{M2} = D_2),$$

there does not exist a single sub-state-tree that will have the same set of basic state trees that satisfy P. That is why we use BDD-based predicate computation instead of a sub-state-tree set in the synthesis stage.

Lemma 4.3 Let $\mathbf{G} = (\mathbf{ST}, \mathcal{H}, \Sigma, \Delta, \mathbf{ST}_o, \mathcal{ST}_m)$ be a state tree structure with $\mathbf{ST} = (X, x_o, \mathcal{T}, \mathcal{E})$. Let $b = (X_b, x_o, \mathcal{T}_b, \mathcal{E}_b)$ be a basic sub-state-tree of $\mathbf{ST}$ and $\mathbf{ST}_1 = (X_1, x_o, \mathcal{T}_1, \mathcal{E}_1)$ a sub-state-tree of $\mathbf{ST}$. Then

$$b \models \Theta(\mathbf{ST}_1) \Leftrightarrow b \leq \mathbf{ST}_1.$$

Proof. 1. ($\Leftarrow$). $\Theta(b) \preceq \Theta(\mathbf{ST}_1)$ and $b \models \Theta(b)$ from the item 2 and 3 of Lemma 4.2 implies $b \models \Theta(\mathbf{ST}_1)$.
2. ($\Rightarrow$). Assume $b \models \Theta(\mathbf{ST}_1)$, i.e., $\Theta(\mathbf{ST}_1)(b) = 1$. We prove $X_b \subseteq X_1$ by induction on the level number n.
   a) (base case) Let the level number $n = 0$. Notice that $\mathbf{ST}_1$ cannot be the empty state tree because otherwise $\Theta(\mathbf{ST}_1) := 0$ contradicts $\Theta(\mathbf{ST}_1)(b) = 1$. Then the root state $x_o \in X_b \cap X_1$. Thus $x_o \in X_b \Rightarrow x_o \in X_1$ as required.

b) (inductive case) Assume that for any state x with level number less than k, $x \in X_b \Rightarrow x \in X_1$. Now let $y \in \mathcal{E}_b(x)$ and $\mathcal{LV}(y) = k$ (the level number of y is k). We need to prove $y \in \mathcal{E}_1(x)$ as well. There are two cases:
   i. if $\mathcal{T}(x) = and$, $\mathcal{E}_b(x) = \mathcal{E}_1(x) = \mathcal{E}(x)$ from the definition of sub-state-tree. Thus $y \in \mathcal{E}_1(x)$;
   ii. if $\mathcal{T}(x) = or$, then y is the only child of x on b (b is a basic state tree). Then the term $v_x = y \wedge \Theta(\mathbf{S}T_1^y)$ must appear in $\Theta(\mathbf{S}T_1^x)$ and therefore $y \in \mathcal{E}_1(x)$ as required. Otherwise assigning y, the only child of x on b, to the state variable v_x will give $\Theta(\mathbf{S}T_1^x)(b) = 0$. Then going up along the state tree b, for any ancestor z of x, including the root state x_o, we have $\Theta(\mathbf{S}T_1^z)(b) = 0$ as well. This contradicts $\Theta(\mathbf{S}T_1)(b) = 1$.

Finally, we can claim $(\forall x \in X_b) x \in X_1$. Then $b \leq \mathbf{S}T_1$ as required.

Combining Lemma 4.3 and item 2 of Lemma 4.2, we have

$$b \models \Theta(\mathbf{S}T_1) \Leftrightarrow b \leq \mathbf{S}T_1 \Leftrightarrow \Theta(b) \preceq \Theta(\mathbf{S}T_1).$$

Now we can give an important proposition.

Proposition 4.1 Let $\mathbf{S}T_1$ be a sub-state-tree of $\mathbf{S}T$. $\Theta(\mathbf{S}T_1)$ is the characteristic function, or predicate of the set $\mathcal{B}(\mathbf{S}T_1)$.

Proof. We need to prove that $b \in \mathcal{B}(\mathbf{S}T_1) \Leftrightarrow b \models \Theta(\mathbf{S}T_1)$. This follows directly from Lemma 4.3.

Proposition 4.2 Let $\{\mathbf{S}T_i | i \in I\}$, for index set I, be a set of sub-state-trees of $\mathbf{S}T$. Then $\bigvee_{i \in I} \Theta(\mathbf{S}T_i)$ is the characteristic function of the set $\bigcup_{i \in I} \mathcal{B}(\mathbf{S}T_i)$.

Proof. We need to prove that $b \in \bigcup_{i \in I} \mathcal{B}(\mathbf{S}T_i) \Leftrightarrow b \models \bigvee_{i \in I} \Theta(\mathbf{S}T_i)$.
1. ($\Rightarrow$). Let $b \in \bigcup_{i \in I} \mathcal{B}(\mathbf{S}T_i)$. Then $b \in \mathcal{B}(\mathbf{S}T_i)$ for some i. By Proposition 4.1, $b \models \Theta(\mathbf{S}T_i) \preceq \bigvee_{i \in I} \Theta(\mathbf{S}T_i)$ So $b \models \bigvee_{i \in I} \Theta(\mathbf{S}T_i)$ as required.
2. ($\Leftarrow$). Reverse the steps in the proof of ($\Rightarrow$).

Proposition 4.2 tells us that an arbitrary set B_P of basic state trees has its characteristic function written by

$$P := \bigvee_{b \in B_P} \Theta(b).$$

After the encoding, the intersection, union and complement operations over sets can be equivalently performed respectively by the meet, join and negation operators over predicates. We demonstrate these operations in the following example.

Example 4.1 As illustrated in Figure 4.8, there are 3 OR superstates on $\mathbf{S}T$, so we bring in 3 state variables, $v_{x_1}, v_{x_2}, v_{x_{22}}$, for the OR superstates x_1, x_2 and x_{22}, respectively.

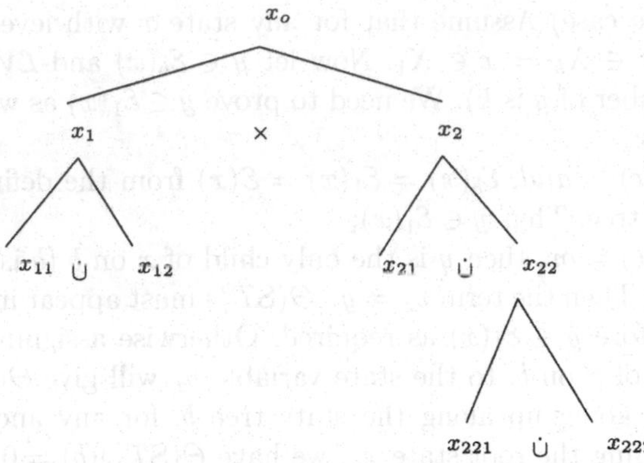

Fig. 4.8. State tree **ST**

1. Set intersection.
   By Lemma 2.9, the intersection of two sets of basic state trees, i.e., $\mathcal{B}(\mathbf{ST}_1) \cap \mathcal{B}(\mathbf{ST}_2)$, is equivalent to the meet of two sub-state-trees, i.e., $\mathbf{ST}_1 \wedge \mathbf{ST}_2$. So we just need to demonstrate that $\Theta(\mathbf{ST}_1 \wedge \mathbf{ST}_2) \equiv \Theta(\mathbf{ST}_1) \wedge \Theta(\mathbf{ST}_2)$. An example is shown in Figure 4.9. We have
$$\begin{aligned}\Theta(\mathbf{ST}_1) \wedge \Theta(\mathbf{ST}_2) := & (v_{x_1} = x_{11} \wedge v_{x_2} = x_{22}) \wedge \\ & (v_{x_2} = x_{21} \vee v_{x_2} = x_{22} \wedge v_{x_{22}} = x_{221}) \\ \equiv\ & v_{x_1} = x_{11} \wedge v_{x_2} = x_{22} \wedge v_{x_{22}} = x_{221} \\ \equiv\ & \Theta(\mathbf{ST}_3)\end{aligned}$$
   as required.
2. Set union.
   The union of two basic state tree sets does not in general have a state tree representation. So we use the simple example shown in Figure 4.10. $\mathbf{ST}_1$ and $\mathbf{ST}_2$ are basic state trees of $\mathbf{ST}$ (Fig. 4.8). Given two sets of basic state trees as $\{\mathbf{ST}_1\}$ and $\{\mathbf{ST}_2\}$, the characteristic function of the union

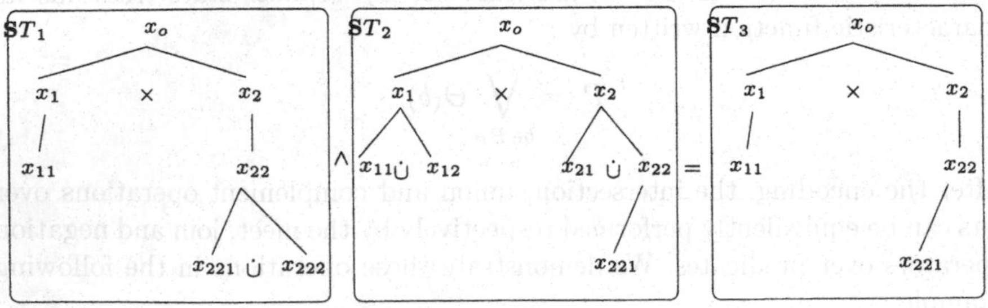

Fig. 4.9. Set intersection

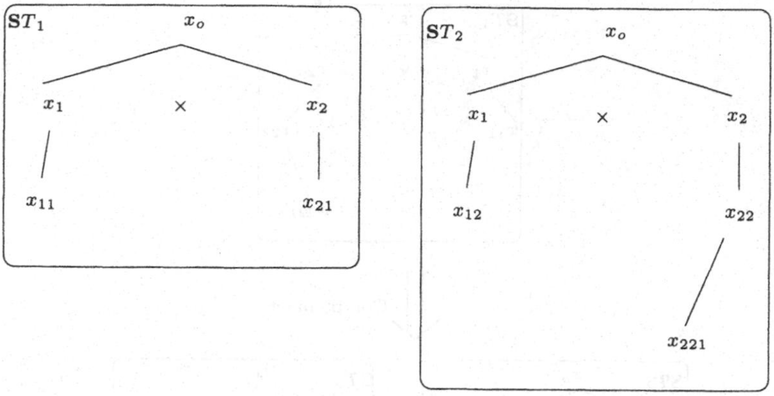

Fig. 4.10. Set union

$\{\mathbf{ST}_1, \mathbf{ST}_2\}$ is obtained by Proposition 4.2.

$$\Theta(\mathbf{ST}_1) \vee \Theta(\mathbf{ST}_2) := (v_{x_1} = x_{11} \wedge v_{x_2} = x_{21}) \vee$$
$$(v_{x_1} = x_{12} \wedge v_{x_2} = x_{22} \wedge v_{x_{22}} = x_{221}).$$

The reader may verify that the above formula $\Theta(\mathbf{ST}_1) \vee \Theta(\mathbf{ST}_2)$ is indeed the characteristic function of the union $\{\mathbf{ST}_1, \mathbf{ST}_2\}$ by checking all basic state trees of $\mathbf{ST}$.

3. Set complement.
By Proposition 4.1, the complement set of $\mathcal{B}(\mathbf{ST}_1)$ has the characteristic function $\neg\Theta(\mathbf{ST}_1)$. In Figure 4.11, $\mathbf{ST}_1$ is a sub-state-tree of $\mathbf{ST}$ (Fig. 4.8). The complement set of $\mathcal{B}(\mathbf{ST}_1)$ is given by $\mathcal{B}(\mathbf{ST}_2) \cup \mathcal{B}(\mathbf{ST}_3)$ (as is easily verified by enumerating $\mathcal{B}(\mathbf{ST}_1)$, $\mathcal{B}(\mathbf{ST}_2)$ and $\mathcal{B}(\mathbf{ST}_3)$). We need to demonstrate that $\neg\Theta(\mathbf{ST}_1)$ is indeed the characteristic function of $(\mathcal{B}(\mathbf{ST}_2) \cup \mathcal{B}(\mathbf{ST}_3))$, i.e., $\neg\Theta(\mathbf{ST}_1) \equiv \Theta(\mathbf{ST}_2) \vee \Theta(\mathbf{ST}_3)$. First we compute

$$\Theta(\mathbf{ST}_2) \vee \Theta(\mathbf{ST}_3) := (v_{x_1} = x_{12}) \vee$$
$$(v_{x_1} = x_{11} \wedge (v_{x_2} = x_{21} \vee v_{x_2} = x_{22} \wedge v_{x_{22}} = x_{222})).$$

Then compute $\neg\Theta(\mathbf{ST}_1)$ from $\Theta(\mathbf{ST}_1)$ by

$$\begin{aligned}\neg\Theta(\mathbf{ST}_1) &:= \neg(v_{x_1} = x_{11} \wedge (v_{x_2} = x_{22} \wedge v_{x_{22}} = x_{221}))\\ &\equiv \neg(v_{x_1} = x_{11}) \vee \neg(v_{x_2} = x_{22} \wedge v_{x_{22}} = x_{221})\\ &\equiv v_{x_1} = x_{12} \vee (v_{x_2} = x_{21} \vee v_{x_{22}} = x_{222}).\end{aligned}$$

Notice that $\neg(v_{x_1} = x_{11}) \equiv (v_{x_1} = x_{12})$ because the range of v_{x_1} is $\{x_{11}, x_{12}\}$.
This does not look like $\Theta(\mathbf{ST}_2) \vee \Theta(\mathbf{ST}_3)$. But investigating it carefully one finds

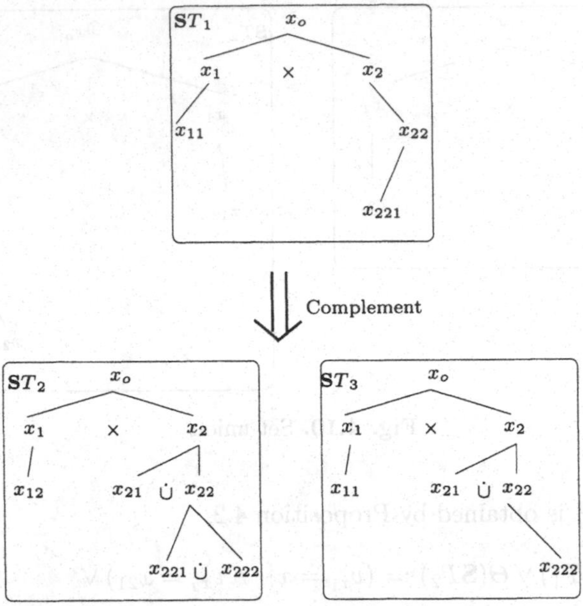

Fig. 4.11. Set complement

$$
\begin{aligned}
\neg\Theta(\mathbf{ST}_1) &\equiv v_{x_1} = x_{12} \vee (v_{x_2} = x_{21} \vee v_{x_{22}} = x_{222} \wedge 1) \\
&\equiv v_{x_1} = x_{12} \vee (v_{x_2} = x_{21} \vee v_{x_{22}} = x_{222} \wedge (v_{x_2} = x_{21} \vee v_{x_2} = x_{22})) \\
&\equiv v_{x_1} = x_{12} \vee (v_{x_2} = x_{21} \vee v_{x_{22}} = x_{222} \wedge v_{x_2} = x_{22}) \\
&\equiv v_{x_1} = x_{12} \vee (v_{x_2} = x_{21} \vee v_{x_{22}} = x_{222} \wedge v_{x_2} = x_{22}) \wedge 1 \\
&\equiv v_{x_1} = x_{12} \vee (v_{x_2} = x_{21} \vee v_{x_{22}} = x_{222} \wedge v_{x_2} = x_{22}) \wedge \\
&\qquad (v_{x_1} = x_{11} \vee v_{x_1} = x_{12}) \\
&\equiv (v_{x_1} = x_{12}) \vee (v_{x_1} = x_{11} \wedge (v_{x_2} = x_{21} \vee v_{x_2} = x_{22} \wedge v_{x_{22}} = x_{222})) \\
&\equiv \Theta(\mathbf{ST}_2) \vee \Theta(\mathbf{ST}_3)
\end{aligned}
$$

as required.

Now given a specification $\mathcal{S}$, a set of illegal sub-state-trees, we can write the *illegal predicate* P by

$$P := \bigvee_{S \in \mathcal{S}} \Theta(S).$$

We do not need to go back from a predicate to a set of sub-state-trees in our setting. The controller implementation can be derived directly based on the controllable and coreachable predicate which we will address later in this chapter. However, it is not so difficult to visualize such an inverse function. The problem is that computing such a function can be very difficult, simply because the size of the resulting set of basic state trees can be huge compared to the length of the predicate formula.

4.2.2 Representation of Transitions by Predicates

As described in Chapter 2, a transition $q = \delta_I^x(p, \sigma)$ is extended to $\mathbf{ST}_2^x = \bar{\delta}_I^x(\mathbf{ST}_1^x, \sigma)$. Suppose $\mathbf{ST}_1$ is a sub-state-tree having a child state tree $\mathbf{ST}_1^x$ (with root state x). Then after the transition is triggered, the child state tree $\mathbf{ST}_1^x$ is replaced by $\mathbf{ST}_2^x$ on $\mathbf{ST}_1$ such that $\mathbf{ST}_1$ is transformed to another sub-state-tree. Now look at this transition from the logic point of view. This transition means that a predicate formula $\Theta(\mathbf{ST}_1^x)$ is replaced by $\Theta(\mathbf{ST}_2^x)$ in $\Theta(\mathbf{ST}_1)$ after the transition is executed.

Let $\mathbf{G} = (\mathbf{ST}, \mathcal{H}, \Sigma, \Delta, \mathbf{ST}_o, \mathcal{ST}_m)$ be a state tree structure with $\mathbf{ST} = (X, x_o, \mathcal{T}, \mathcal{E})$. In this section, the objective is to build such a transition relation for each event $\sigma \in \Sigma$ on $\mathbf{G}$. This is in line with our philosophy that no *global* transition relation is built during the synthesis.

For each OR superstate x of $\mathbf{G}$, denote by v_x (v_x') its target (source) state variable, and by $\mathbf{v}$ ($\mathbf{v}'$) the set of all target (source) state variables on $\mathbf{G}$. Call v_x the *normal* state variable of x and v_x' the *prime* state variable of x. The range of both variables is the set of all child states of x. Denote by $P(\mathbf{v}') := P(\mathbf{v})[\mathbf{v} \to \mathbf{v}']$ the predicate defined over $\mathbf{v}'$ by replacing each variable v_x by its prime variable v_x' in $P(\mathbf{v})$. Then in holon H^x, the *local transition relation* for $T = \bar{\delta}_I^x(S, \sigma)$, denoted by transition $\mathbf{t}$, is encoded simply by

$$\Theta(S)[\mathbf{v_{t,S}} \to \mathbf{v_{t,S}}'] \wedge \Theta(T),$$

where $\mathbf{v_{t,S}}$ and $\mathbf{v_{t,T}}$ are the sets of variables appearing in $\Theta(S)$ and $\Theta(T)$, respectively. For example, the transition labelled by event α_1 of the Small Factory in Figure 4.6 of page 84 is represented by

$$v_{M1}' = I_1 \wedge v_{M1} = W_1.$$

The transition $\mathbf{t}$ ($T = \bar{\delta}_I^x(S, \sigma)$) can occur only when the system is staying inside the superstate x. So the local transition relation is not enough to completely describe the transition. Thus, we define the *transition relation* $N_\mathbf{t}$ for $\mathbf{t}$ by [5]

$$N_\mathbf{t} := \Theta(S)[\mathbf{v_{t,S}} \to \mathbf{v_{t,S}}'] \wedge \Theta(\widehat{T}).$$

The sub-state-tree $\widehat{T}$ of $\mathbf{ST}$ must have the child state tree T. Because it must also make sure the system is at superstate x when the transition $\mathbf{t}$ is triggered (because σ occurs inside x), $\widehat{T}$ should not have any states that are *exclusive* from x [6]. Formally, a state $a \in \widehat{T}$ if and only if

[5] As our synthesis only requires the computation of Γ symbolically, we define $N_\mathbf{t}$ like this. One can also define $N_\mathbf{t} := \Theta(\widehat{S}) \wedge \Theta(T)[\mathbf{v_{t,T}} \to \mathbf{v_{t,T}}']$ if it is desired to use $N_\mathbf{t}$ to compute Δ symbolically. Also, compared with the transition relations defined in symbolic model checking, our encoding is more economical, because we omit redundant terms like $v_y = v_y'$.

[6] Refer to page 18 of Chapter 2 for the concept of *exclusive*.

1. $a|x$, or
2. $a \le x$, or
3. $x \le a \Rightarrow a \in T$.

Now the transition $\mathbf{t}$ is completely decided by the triple $(N_{\mathbf{t}}, \mathbf{v_{t,S}}, \mathbf{v_{t,T}})$. $\mathbf{v_{t,S}}$ and $\mathbf{v_{t,T}}$ are the sets of variables appearing in $\Theta(S)$ and $\Theta(T)$, respectively. It is well to note that $\mathbf{v_{t,T}}$ is the set of state variables for $\Theta(T)$ instead of $\Theta(\widehat{T})$, because during the computation of Γ, the change of values can only happen inside $\mathbf{v_{t,S}}$ and $\mathbf{v_{t,T}}$ upon the occurrence of transition $\mathbf{t}$ and any other state variables will keep their original values. Also keep in mind that this encoding *only* applies to well-formed STS that respect the *local coupling* property. [7]

An event σ can occur at several states of a holon sequentially and inside a set of holons concurrently. Let D_σ be the set of OR superstates where σ labels some of the inner transitions of their matched holons. Let T_σ^x be the set of inner transitions in holon H^x that are labelled by σ. Then the *symbolic representation* of the entire set of transitions labelled by *a given event* σ is given as a triple $(N_\sigma, \mathbf{v}_{\sigma,\mathbf{S}}, \mathbf{v}_{\sigma,\mathbf{T}})$ where

1. $N_\sigma := \bigwedge_{x\in D_\sigma} \bigvee_{\mathbf{t}\in T_\sigma^x} N_{\mathbf{t}}$;
2. $\mathbf{v}_{\sigma,\mathbf{S}} := \bigcup_{x\in D_\sigma} \bigcup_{\mathbf{t}\in T_\sigma^x} \mathbf{v_{t,S}}$;
3. $\mathbf{v}_{\sigma,\mathbf{T}} := \bigcup_{x\in D_\sigma} \bigcup_{\mathbf{t}\in T_\sigma^x} \mathbf{v_{t,T}}$.

Notice that the local coupling requirement is essential in order to give such a simple definition of the triple. Also notice that because events are shared locally in a well-defined STS, the variables appearing in N_σ can only be

1. the normal state variables for all of the OR ancestors of the superstates in D_σ, and
2. the normal state variables in $\mathbf{v}_{\sigma,\mathbf{T}}$ for describing the target states, and
3. the prime state variables in $\mathbf{v}_{\sigma,\mathbf{S}}'$ for describing the source states.

The example in Figure 4.12 illustrates how to write transitions symbolically. The event set is $\{\alpha, \beta, \mu, \lambda, \sigma\}$. We show how to obtain the transitions for the first three events.

1. Compute $(N_\mu, \mathbf{v}_{\mu,\mathbf{S}}, \mathbf{v}_{\mu,\mathbf{T}})$.
   μ occurs in H^{b_1}. Denote $\mathbf{ST}_2^{b_1} = \overline{\delta_I^{b_1}}(\mathbf{ST}_1^{b_1}, \mu)$. $\mathbf{ST}_1^{b_1}$ is represented by its active state set $\{b_{10}\}$. So $\Theta(\mathbf{ST}_1^{b_1})[v_{b_1} \to v_{b_1}'] := (v_{b_1}' = b_{10})$. Also $\Theta(\widehat{\mathbf{ST}_2^{b_1}}) := (v_{x_1} = b \wedge v_{b_1} = b_{11})$. Thus,

$$N_\mu := (v_{b_1}' = b_{10}) \wedge (v_{x_1} = b \wedge v_{b_1} = b_{11}).$$

Identifying the prime variables inside b_1, we have

$$\mathbf{v}_{\mu,\mathbf{S}} = \{v_{b_1}\}.$$

[7] Refer to the definition of STS on page 40.

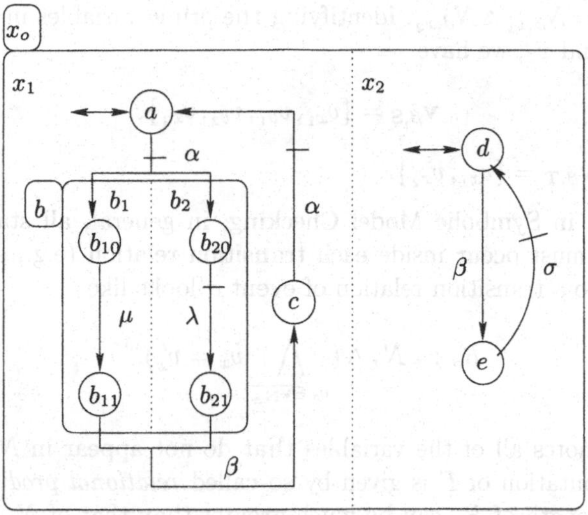

Fig. 4.12. Example: symbolic representation of transitions

Notice that $\mathbf{v}_{\mu,\mathbf{T}} = \{v_{b_1}\}$ even though v_{x_1} also occurs in the formula N_μ. The reason is that $\mathbf{v}_{\mu,\mathbf{T}}$ is defined as the variables inside $\Theta(\mathbf{S}T_2^{b_1})$ instead of $\Theta(\widehat{\mathbf{S}T_2^{b_1}})$.

2. Compute $(N_\alpha, \mathbf{v}_{\alpha,\mathbf{S}}, \mathbf{v}_{\alpha,\mathbf{T}})$.
α only occurs inside the holon H^{x_1} matched to x_1. For the transition from a to superstate b , we write its transition relation as
$$N_{\alpha,1} := (v'_{x_1} = a) \wedge (v_{x_1} = b \wedge v_{b_1} = b_{10} \wedge v_{b_2} = b_{20}).$$
As it can also occur at state c of holon H^{x_1}, another transition relation labelled by α is given by
$$N_{\alpha,2} := (v'_{x_1} = c) \wedge (v_{x_1} = a).$$
Then $N_\alpha := N_{\alpha,1} \vee N_{\alpha,2}$. Identifying the prime variables appearing in N_α that are inside x_1, we have
$$\mathbf{v}_{\alpha,\mathbf{S}} = \{v_{x_1}\}.$$
Similarly, $\mathbf{v}_{\alpha,\mathbf{T}} = \{v_{x_1}, v_{b_1}, v_{b_2}\}$.
3. Compute $(N_\beta, \mathbf{v}_{\beta,\mathbf{S}}, \mathbf{v}_{\beta,\mathbf{T}})$.
β occurs in both H^{x_1} and H^{x_2}. In H^{x_1}, we have
$$N_{\beta,x_1} := (v'_{x_1} = b \wedge v'_{b_1} = b_{11} \wedge v'_{b_2} = b_{21}) \wedge (v_{x_1} = c).$$
In H^{x_2}, we have
$$N_{\beta,x_2} := (v'_{x_2} = d) \wedge (v_{x_2} = e).$$

Then $N_\beta := N_{\beta,x_1} \wedge N_{\beta,x_2}$. Identifying the prime variables in N_β that are inside x_1 and x_2, we have

$$\mathbf{v}_{\beta,\mathbf{S}} = \{v_{x_1}, v_{b_1}, v_{b_2}, v_{x_2}\}.$$

Similarly, $\mathbf{v}_{\beta,\mathbf{T}} = \{v_{x_1}, v_{x_2}\}$.

Recall that in Symbolic Model Checking, in general, all state variables of the system must occur inside each transition relation (e.g., section 2.2.2 of [BCC98]). So a transition relation of event σ looks like

$$R_\sigma := N_\sigma \wedge (\bigwedge_{v_x \in \overline{\mathbf{v}_{\mathbf{N}_\sigma}}} v_x = v'_x),$$

where $\overline{\mathbf{v}_{\mathbf{N}_\sigma}}$ denotes all of the variables that do not appear in N_σ. Then the symbolic computation of Γ is given by so-called *relational product* [BCC98]. However, the length of R_σ can be much greater than that of N_σ for complex systems. Sometimes it is critical to use $(N_\sigma, \mathbf{v}_{\sigma,\mathbf{S}}, \mathbf{v}_{\sigma,\mathbf{T}})$ instead of R_σ as one may not have enough memory in the computer to store R_σ. So by just building $(N_\sigma, \mathbf{v}_{\sigma,\mathbf{S}}, \mathbf{v}_{\sigma,\mathbf{T}})$ instead of R_σ, we gain more in memory saving. It saves time too as Γ defined by $(N_\sigma, \mathbf{v}_{\sigma,\mathbf{S}}, \mathbf{v}_{\sigma,\mathbf{T}})$ needs less computation.

4.3 Symbolic Synthesis of STS

After defining Θ, we can rewrite a specification by its logic formula. Now the remaining job is to compute $\sup \mathcal{C}^2\mathcal{P}(\cdot)$ symbolically. Recalling the algorithm given in Chapter 3, the following algorithm is used to compute $\sup \mathcal{C}^2\mathcal{P}(P)$.

1. Let $K_o := P$.
2. $K_{i+1} := \Omega_P(K_i) := P \wedge CR(\mathbf{G}, \neg[\neg K_i])$.
3. If $K_{i+1} \equiv K_i$, then $\sup \mathcal{C}^2\mathcal{P}(P) := K_i$. Otherwise go back to step 2.

The efficiency of the above algorithm depends on the fast computation of $[\cdot]$ and $CR(\mathbf{G}, \cdot)$.

In this section, we first show how to compute Γ symbolically and then give algorithms for structured computation of $[\cdot]$ and $CR(\mathbf{G}, \cdot)$, respectively.

4.3.1 Computation of Γ

A predicate R is equivalent to a set of sub-state-trees. For each sub-state-tree $T \models R$, we can compute $T' = \Gamma(T, \sigma)$ as the sub-state-tree that is transformed to a sub-state-tree of T if σ is triggered. So there must exist a predicate P that is identified by the set of sub-state-trees defined by

$$\{T' | (\exists T \models R) T' = \Gamma(T, \sigma)\}.$$

What if we define a function $\hat{\Gamma}$ to compute P directly from $P := \hat{\Gamma}(R, \sigma)$? That must speed up our synthesis process as $\hat{\Gamma}$ can be considered as an extended function of Γ that can compute *a set of* sub-state-trees in one call. In a word, $\hat{\Gamma}(R, \sigma)$ will find every sub-state-tree that is transformed into a sub-state-tree satisfying R by some transition labelled with σ, illustrated in Figure 4.13. Notice that in general $R' \subseteq R$. This means that some sub-state-trees in R may not be reached by a transition labelled with σ.

Definition 4.2 [$\exists v_x P$, $\forall v_x P$] Let v_x be a state variable appearing in the predicate P. The range of v_x is $\mathcal{E}(x) = \{y_1, y_2, \cdots, y_n\}$. Denote by $P[y_i/v_x]$ the resulting predicate after assigning y_i to v_x. Then define

$$\exists v_x P := \bigvee_{i=1}^{n} P[y_i/v_x].$$

$$\forall v_x P := \bigwedge_{i=1}^{n} P[y_i/v_x].$$

◇

Remarks

1. The above definition is taken from *Quantifier Elimination* theory [Arn88]. If the domain of a variable is finite, the $\exists$ and $\forall$ operators for that variable can be *eliminated* using the above formulas. This is important because BDD cannot represent first order logic formulas directly.
2. The variable v_x will not appear in $\exists v_x P$ and $\forall v_x P$ as it has been bound (assigned values).
3. The following is an important fact about the two operators:

   $$\forall v_x P \preceq P \preceq \exists v_x P.$$

   The proof is directly from the definition. For example, to prove $\forall v_x P \preceq P$, one just needs to show $b \models \forall v_x P \Rightarrow b \models P$.

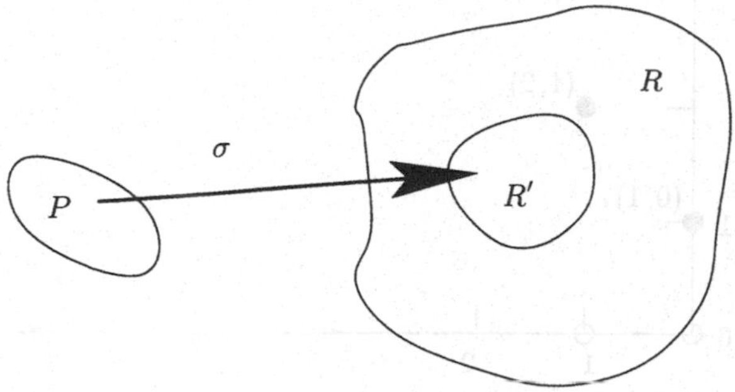

Fig. 4.13. $P = \hat{\Gamma}(R, \sigma)$

4. As found by researchers in Symbolic Model Checking, $\exists v_x P$ will be used to define $\hat{\Gamma}$. For example, let

$$N_\sigma := (v'_x = 0 \wedge v_x = 1) \vee (v'_x = 1 \wedge v_x = 2)$$

be a transition relation for event σ. This means that the transitions labelled by σ transfer state 0 to 1, state 1 to 2. Let $\mathcal{E}(x) = \{0, 1, 2\}$. Then

$$\begin{aligned} \exists v_x N_\sigma &:= N_\sigma[0/v_x] \vee N_\sigma[1/v_x] \vee N_\sigma[2/v_x] \\ &\equiv \mathit{false} \vee (v'_x = 0) \vee (v'_x = 1) \\ &\equiv (v'_x = 0) \vee (v'_x = 1) \end{aligned}$$

will be identified by the source states of σ, $\{0, 1\}$. Thus $\exists v_x P$ is used to find all the source states (written with prime variables) in P.
We provide another description of $\exists$. N_σ can be looked as a set of points $\{(0, 1), (1, 2)\}$ at the $v_{x'} - v_x$ plane shown in Figure 4.14. The $\exists v_x$ is like a *natural projection* from the $v_x - v'_x$ plane to the v'_x axis.
Just as a predicate is the symbol of a set of basic state trees, the operator $\exists$ transforming a predicate to another one is equivalent to transforming its corresponding set to another set. The example in Figure 4.15 demonstrates this. After $\exists v_{x_3} \Theta(\mathbf{ST}_1)$, the sub-state-tree $\mathbf{ST}_1$ in (b) (of $\mathbf{ST}$ in (a)) is transformed to the state tree $\mathbf{ST}_2$ in (c). All descendants of x_3 are added on $\mathbf{ST}_2$ because of the $\exists$ operator. This is confirmed by its symbolic computation as follows.

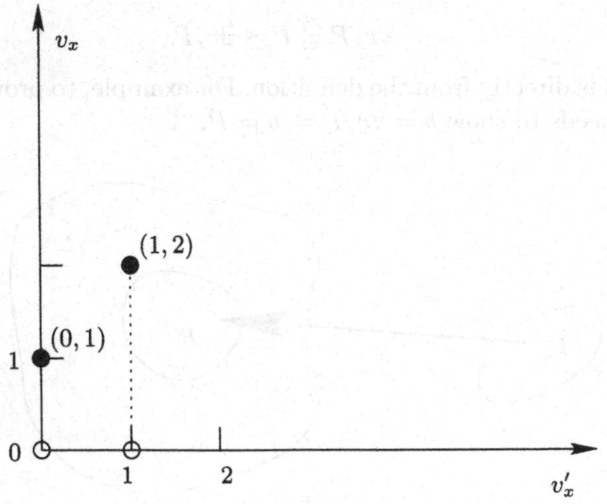

Fig. 4.14. Graphical explanation of $\exists v_x N_\sigma$

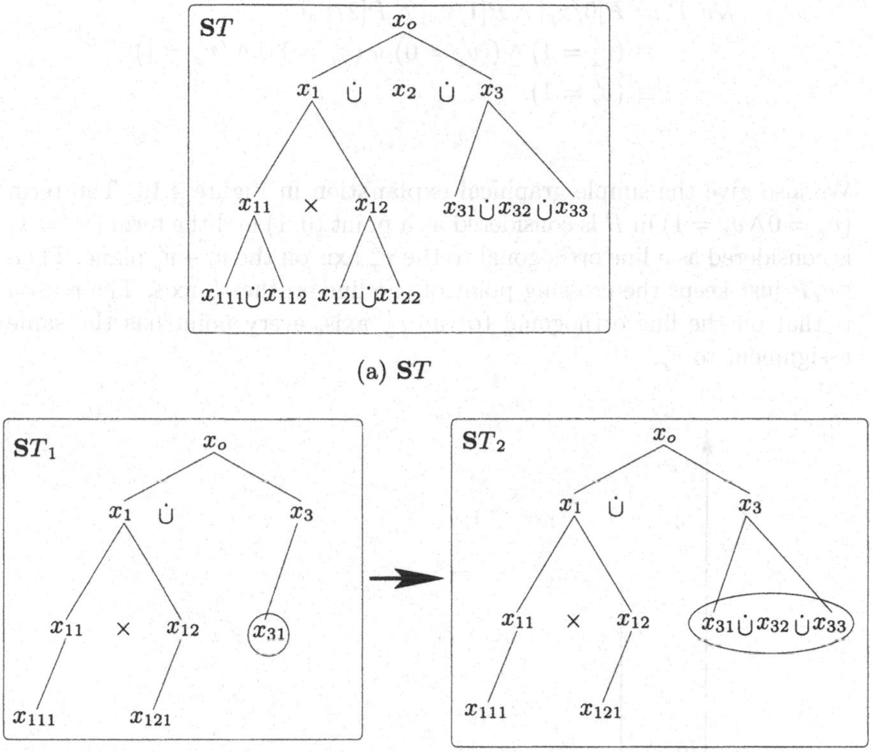

(a) **ST**

(b) **ST**$_1$ (c) **ST**$_2$

Fig. 4.15. Another Explanation of $\exists P$

$$\begin{aligned}
\Theta(\mathbf{ST}_1) &:= (v_{x_o} = x_1 \wedge v_{x_{11}} = x_{111} \wedge v_{x_{12}} = x_{121}) \vee \\
&\quad\ (v_{x_o} = x_3 \wedge v_{x_3} = x_{31}) \\
\exists v_{x_3}\Theta(\mathbf{ST}_1) &:= \Theta(\mathbf{ST}_1)[x_{31}/v_{x_3}] \vee \Theta(\mathbf{ST}_1)[x_{32}/v_{x_3}] \vee \Theta(\mathbf{ST}_1)[x_{33}/v_{x_3}] \\
&\equiv (v_{x_o} = x_1 \wedge v_{x_{11}} = x_{111} \wedge v_{x_{12}} = x_{121}) \vee v_{x_o} = x_3 \\
&\equiv \Theta(\mathbf{ST}_2)
\end{aligned}$$

5. $\forall v_x P$ is the sub-predicate of P that agrees on all the assignments to v_x. So if we know a sub-state-tree $\mathbf{ST}_1 \models \forall v_x P$, we can *infer* that for the sub-state-tree $\mathbf{ST}_2$ obtained by replacing $\mathbf{ST}_1$'s child state tree $\mathbf{ST}_1^x$ with $\mathbf{ST}^x$ (adding all missing branches under x), $\mathbf{ST}_2 \models \forall v_x P$ too. Notice that $\mathbf{ST}_1 \leq \mathbf{ST}_2$. Such *inference* is quite useful in speeding up the synthesis. This $\forall$ operator will be used in the computation of $CR(\mathbf{G}, \cdot)$. Here we give an example. Let

$$P := (v'_x = 0 \wedge v_x = 1) \vee (v'_x = 1)$$

with both the range of v'_x and v_x to be $\{0, 1, 2\}$. Then

$$\begin{aligned}\forall v_x P &:= P[0/v_x] \wedge P[1/v_x] \wedge P[2/v_x] \\ &\equiv (v'_x = 1) \wedge ((v'_x = 0) \vee (v'_x = 1)) \wedge (v'_x = 1) \\ &\equiv (v'_x = 1).\end{aligned}$$

We also give the simple graphical explanation in Figure 4.16. The term $(v'_x = 0 \wedge v_x = 1)$ in P is considered as a point $(0, 1)$ and the term $(v'_x = 1)$ is considered as a line orthogonal to the v'_x axis on the $v_x - v'_x$ plane. Then $\forall v_x P$ just keeps the crossing point of the line on the v'_x axis. The reason is that on the line orthogonal to the v'_x axis, every point has the same assignment to v'_x.

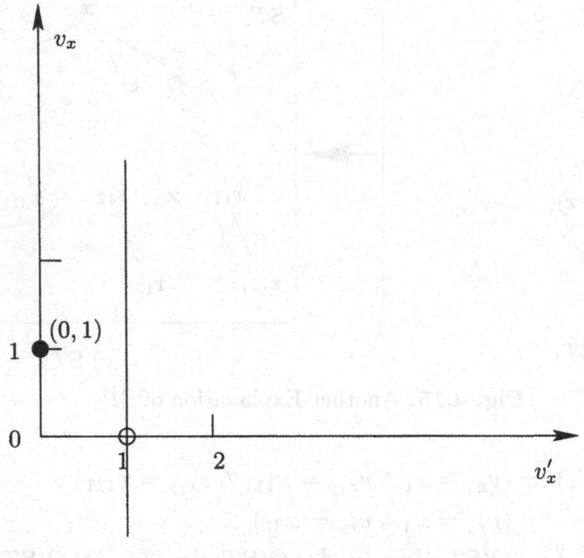

Fig. 4.16. Graphical explanation of $\forall v_x P$

Let $\mathbf{v} = \{v_i | i = 1, 2, \ldots, n\}$ be a set of state variables appearing in P. For convenience, denote

$$\exists \mathbf{v} P := \exists v_1 (\exists v_2 \ldots (\exists v_n P));$$

$$\forall \mathbf{v} P := \forall v_1 (\forall v_2 \ldots (\forall v_n P)).$$

Now we can define $\hat{\Gamma}$ as follows.

Definition 4.3 $[\hat{\Gamma}]$ Let $\mathbf{G} = (\mathbf{ST}, \mathcal{H}, \Sigma, \Delta, \mathbf{ST}_o, \mathcal{ST}_m)$ be a state tree structure with $\mathbf{ST} = (X, x_o, \mathcal{T}, \mathcal{E})$. Let $\sigma \in \Sigma$. Let $(N_\sigma, \mathbf{v}_{\sigma,\mathbf{S}}, \mathbf{v}_{\sigma,\mathbf{T}})$ represent the entire set of transitions labelled by σ. Let $Pred(\mathbf{ST})$ be defined over the normal variable set $\mathbf{v}$. Then $\hat{\Gamma} : Pred(\mathbf{ST}) \times \Sigma \rightarrow Pred(\mathbf{ST})$ is defined by

$$\hat{\Gamma}(P, \sigma) := (\exists \mathbf{v}_{\sigma,\mathbf{T}}(P \wedge N_\sigma))[\mathbf{v}_{\sigma,\mathbf{S}}' \rightarrow \mathbf{v}_{\sigma,\mathbf{S}}].$$

We can omit ^ and will write $\Gamma(P, \sigma)$ without ambiguity.

◊

Remarks

1. This definition is inspired by the *relational product* [BCC98] concept that is used extensively in symbolic model checking with BDD. Our definition requires less computation than relational product, but is only applied to well-formed STS. For example, if events are shared among different levels, the above definition will no longer be valid.
2. To compute $\Gamma(P, \sigma)$, first compute $P \wedge N_\sigma$. The resulting predicate has the transitions of σ with their target state trees satisfying P. [8] Then we quantify out variables in $\mathbf{v}_{\sigma,\mathbf{T}}$ by the $\exists$ operator to get the source sub-state-trees. The resulting predicate has prime variables in $\mathbf{v}_{\sigma,\mathbf{S}}'$ (because N_σ is defined over prime variable set $\mathbf{v}_{\sigma,\mathbf{S}}'$). So we replace them by normal variables such that the final predicate is defined over the normal variable set $\mathbf{v}$ again.
3. $\Gamma(P, \sigma)$ returns the predicate that can reach P via transition σ. If $P \equiv \Theta(\mathbf{ST}_1)$, then it will give the same result as $\Gamma(\mathbf{ST}_1, \sigma)$ defined in chapter 2. Comparing both definitions carefully, one can verify that they have the same functionality.
4. We use the same STS model as shown in Figure 4.12 to demonstrate how to compute Γ. For convenience, we draw the same graph again here in Figure 4.17. Let $P := v_{x_1} = b \vee v_{x_1} = c$. P is the predicate representation

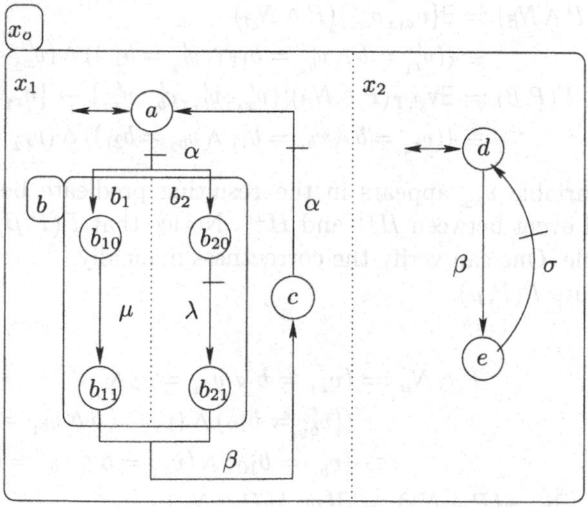

Fig. 4.17. Example: Γ

[8] Because P is written by variable set $\mathbf{v}$ instead of $\mathbf{v}'$, it is identified as a set of target sub-state-trees instead of source sub-state-trees.

of a sub-state-tree that has active state set $\{b, c\}$.

a) Compute $\Gamma(P, \alpha)$.

$$\begin{aligned} P \wedge N_\alpha := & (v_{x_1} = b \vee v_{x_1} = c) \wedge \\ & ((v'_{x_1} = a) \wedge (v_{x_1} = b \wedge v_{b_1} = b_{10} \wedge v_{b_2} = b_{20}) \vee \\ & (v'_{x_1} = c) \wedge (v_{x_1} = a)) \\ \equiv & (v'_{x_1} = a) \wedge (v_{x_1} = b \wedge v_{b_1} = b_{10} \wedge v_{b_2} = b_{20}). \\ \exists \mathbf{v}_{\alpha,\mathbf{T}}(P \wedge N_\alpha) := & \exists\{v_{x_1}, v_{b_1}, v_{b_2}\}(P \wedge N_\alpha) \\ \equiv & v'_{x_1} = a. \end{aligned}$$

Then $\Gamma(P, \alpha) := (v'_{x_1} = a)[v'_{x_1} \to v_{x_1}] \equiv (v_{x_1} = a)$. This means that the sub-state-tree that can reach P by event α is the one that has active state set $\{a\}$. Looking into this system, one can verify that it is indeed the case because α is local inside x_1 and there is a transition labelled by α from a to b.

b) Compute $\Gamma(P, \beta)$.

$$\begin{aligned} P \wedge N_\beta := & (v_{x_1} = b \vee v_{x_1} = c) \wedge \\ & [((v'_{x_1} = b \wedge v'_{b_1} = b_{11} \wedge v'_{b_2} = b_{21}) \wedge (v_{x_1} = c)) \wedge \\ & ((v'_{x_2} = d) \wedge (v_{x_2} = e))] \\ \equiv & ((v'_{x_1} = b \wedge v'_{b_1} = b_{11} \wedge v'_{b_2} = b_{21}) \wedge (v_{x_1} = c)) \wedge \\ & ((v'_{x_2} = d) \wedge (v_{x_2} = e)). \\ \exists \mathbf{v}_{\beta,\mathbf{T}}(P \wedge N_\beta) := & \exists\{v_{x_1}, v_{x_2}\}(P \wedge N_\beta) \\ \equiv & ((v'_{x_1} = b \wedge v'_{b_1} = b_{11} \wedge v'_{b_2} = b_{21})) \wedge (v'_{x_2} = d). \\ \Gamma(P, \beta) := & \exists \mathbf{v}_{\beta,\mathbf{T}}(P \wedge N_\beta)[\{v'_{x_1}, v'_{b_1}, v'_{b_2}, v'_{x_2}\} \to \{v_{x_1}, v_{b_1}, v_{b_2}, v_{x_2}\}] \\ \equiv & ((v_{x_1} = b \wedge v_{b_1} = b_{11} \wedge v_{b_2} = b_{21})) \wedge (v_{x_2} = d). \end{aligned}$$

The variable v_{x_2} appears in the resulting predicate because β is a shared event between H^{x_1} and H^{x_2}. Notice that $\Gamma(P, \beta) \preceq P$ in this example. One can verify the correctness manually.

c) Compute $\Gamma(P, \mu)$.

$$\begin{aligned} P \wedge N_\mu := & (v_{x_1} = b \vee v_{x_1} = c) \wedge \\ & ((v'_{b_1} = b_{10}) \wedge (v_{x_1} = b \wedge v_{b_1} = b_{11})) \\ \equiv & (v'_{b_1} = b_{10}) \wedge (v_{x_1} = b \wedge v_{b_1} = b_{11}). \\ \exists \mathbf{v}_{\mu,\mathbf{T}}(P \wedge N_\beta) := & \exists\{v_{b_1}\}(P \wedge N_\mu) \\ \equiv & (v'_{b_1} = b_{10}) \wedge (v_{x_1} = b). \\ \Gamma(P, \mu) := & ((v'_{b_1} = b_{10}) \wedge (v_{x_1} = b))[v'_{b_1} \to v_{b_1}] \\ \equiv & (v_{b_1} = b_{10}) \wedge (v_{x_1} = b). \end{aligned}$$

$\Gamma(P,\mu) \preceq P$ in this example. The reason is that μ is a local event inside b_1 and all sub-state-trees with b_1 already satisfy P. This is an important observation because it provides a condition under which we know $\Gamma(P,\mu) \preceq P$ without computing $\Gamma(P,\mu)$! In fact, the condition can be given as a simple formula

$$\forall v_{b_1} P \equiv P.$$

Notice that $\forall v_{b_1} P \preceq P$ in general. So the condition means $P \preceq \forall v_{b_1} P$, i.e., $\mathbf{ST}_1 \models P \Rightarrow \mathbf{ST}_2 \models P$, where $\mathbf{ST}_2$ is obtained by replacing $\mathbf{ST}_1$'s child state tree $\mathbf{ST}_1^{b_1}$ with $\mathbf{ST}^{b_1}$ (adding all missing branches under b_1). That is, $\Theta(\mathbf{ST}_2) \preceq P$. As we know the transition μ cannot change any other state variables except v_{b_1},

$$(\forall \mathbf{ST}_1 \models P)\Gamma(\mathbf{ST}_1,\mu) \preceq \Theta(\mathbf{ST}_2).$$

Thus $\Gamma(P,\mu) \preceq P$ from the above condition. We will take advantage of this observation in the computation of $CR(\mathbf{G},\cdot)$.

4.3.2 Computation of $[\cdot]$

From the definition of $[\cdot]$, we can compute $[R]$ from the basic algorithm given as follows.

Algorithm 4.1 [Basic algorithm for $[\cdot]$]
$[R]$

1. $K_o := R$.
2. $K_{i+1} := K_i \vee (\bigvee_{\sigma_u \in \Sigma_u} \Gamma(K_i, \sigma_u))$.
3. If $K_{n+1} \equiv K_n$, then $[R] := K_n$. Otherwise go back to step 2.

$\diamond$

The algorithm 4.1 must terminate in a finite number of steps as $Pred(\mathbf{ST})$ is finite. A simple example is illustrated in Figure 4.18. Let $R := v_x = a$. Then

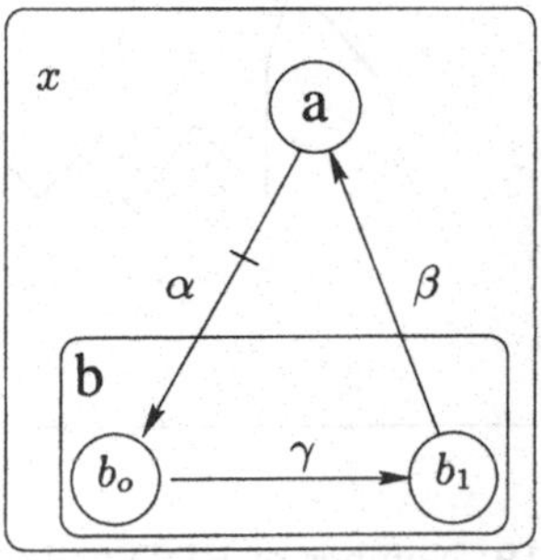

Fig. 4.18. Example for using Algorithm 4.1

$$
\begin{aligned}
K_o &:= v_x = a. \\
K_1 &:= v_x = a \vee (v_x = b \wedge v_b = b_1). \\
K_2 &:= v_x = a \vee (v_x = b \wedge v_b = b_1) \vee \\
&\quad\ (v_x = b \wedge v_b = b_o) \\
&\equiv v_x = a \vee v_x = b \\
&\equiv 1. \\
K_3 &:= 1.
\end{aligned}
$$

So $[R] := 1$. An important observation is that the formula length of intermediate K_i is usually much greater than that of the resulting $[R]$. Thus the biggest disadvantage of this simple algorithm is that the number of BDD nodes for the intermediate predicates $K_i, i = 0, 1, \ldots, n-1$ can be much larger than that of the resulting predicate $[R]$. This is illustrated in Figure 4.19. With the same $[R]$, a better algorithm would have a smaller number of BDD nodes for K_i, i.e., it would save space. It would save time as well because the smaller a BDD, the faster the computation.

To find a better algorithm, there are different techniques that we can use here. One is to find heuristic methods to look for the best order of variables in BDD, which is at present outside our scope. Another one is to take advantage of the structural information of STS.

Structured Computation of $[\cdot]$

As demonstrated before, if x is an OR superstate, then the following tautology will help to simplify the intermediate predicate formula K_i in Algorithm 4.1:

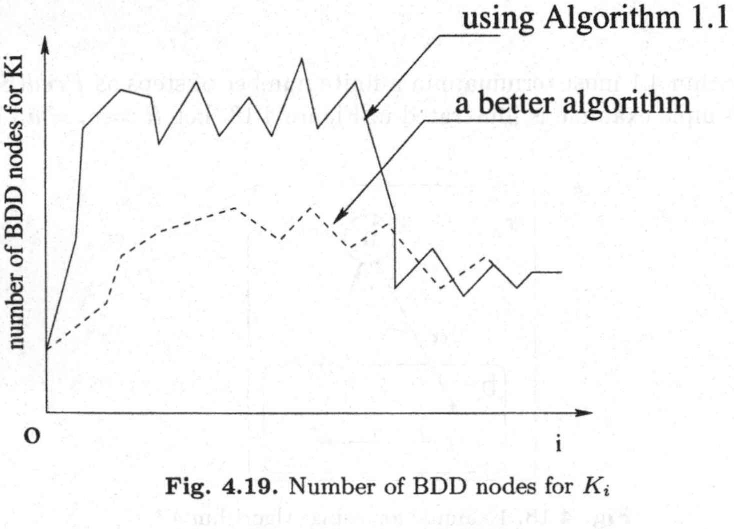

Fig. 4.19. Number of BDD nodes for K_i

$$(\bigvee_{y\in\mathcal{E}(x)} (v_x = y)) \equiv 1,$$

if v_x appears in K_i. This tautology can be applied *only when* the left side of the above tautology appears in K_i. Also notice that to add a term $v_x = y$ in K_i by $\Gamma(K_i, \sigma)$, the event σ must be in Σ^x. So to improve the efficiency of using the above tautology, we need to give a direction of computation based on the system's structure.

The following example in Figure 4.20 demonstrates the above idea. Let

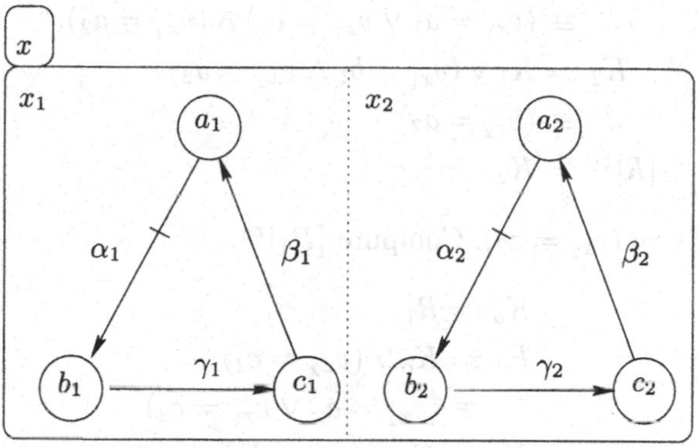

Fig. 4.20. Example for computing $[\cdot]$

$R := v_{x_1} = a_1 \wedge v_{x_2} = a_2$. Using Algorithm 4.1, we have

$K_o := R.$

$$K_1 := K_o \vee \underbrace{(v_{x_1} = c_1 \wedge v_{x_2} = a_2)}_{\text{by } \Gamma(K_o, \beta_1)} \vee \underbrace{(v_{x_1} = a_1 \wedge v_{x_2} = c_2)}_{\text{by } \Gamma(K_o, \beta_2)}$$

$$\equiv (v_{x_1} = a_1 \wedge (v_{x_2} = a_2 \vee v_{x_2} = c_2)) \vee (v_{x_2} = a_2 \wedge (v_{x_1} = a_1 \vee v_{x_1} = c_1)).$$

$$K_2 := K_1 \vee (v_{x_1} = b_1 \wedge v_{x_2} = a_2) \vee (v_{x_1} = c_1 \wedge v_{x_2} = c_2) \vee (v_{x_1} = a_1 \wedge v_{x_2} = b_2)$$

$$\equiv v_{x_1} = a_1 \vee v_{x_2} = a_2 \vee (v_{x_1} = c_1 \wedge v_{x_2} = c_2).$$

$$K_3 := K_2 \vee (v_{x_1} = c_1 \wedge v_{x_2} = b_2) \vee (v_{x_1} = b_1 \wedge v_{x_2} = c_2)$$

$$\equiv (v_{x_1} = a_1 \vee v_{x_1} = c_1) \vee (v_{x_2} = a_2 \vee v_{x_2} = c_2).$$

$$K_4 := K_3 \vee (v_{x_1} = b_1 \wedge v_{x_2} = b_2)$$

$$\equiv (v_{x_1} = a_1 \vee v_{x_1} = c_1 \vee v_{x_1} = b_1) \vee (v_{x_2} = a_2 \vee v_{x_2} = c_2 \vee v_{x_2} = b_2)$$

$$\equiv 1.$$

$[R] := 1.$

Even though $[R] := 1$ is the shortest predicate, the intermediate predicate K_i can be much more complicated.

Now we try another scheme. For each OR superstate x, define $[R]^x$ as the fixpoint of the following iteration

1. $K_o := R$.
2. $K_{i+1} := K_i \vee (\bigvee_{\sigma_u \in \Sigma^x_{Iu}} \Gamma(K_i, \sigma_u))$.
3. If $K_{n+1} \equiv K_n$, then $[R]^x := K_n$. Otherwise go back to step 2.

Σ^x_{Iu} is the internal uncontrollable event set in H^x. Thus $[R]^x$ includes all basic state trees that can reach R by uncontrollable events *occurring in* H^x. For the same example, first compute $[R]^{x_1}$.

$$\begin{aligned}
K_o &:= R. \\
K_1 &:= K_o \vee (v_{x_1} = c_1 \wedge v_{x_2} = a_2) \\
&\equiv (v_{x_1} = a_1 \vee v_{x_1} = c_1) \wedge (v_{x_2} = a_2). \\
K_2 &:= K_1 \vee (v_{x_1} = b_1 \wedge v_{x_2} = a_2) \\
&\equiv (v_{x_2} = a_2). \\
[R]^{x_1} &:= K_2.
\end{aligned}$$

Let $R_1 := [R]^{x_1} := (v_{x_2} = a_2)$. Compute $[R_1]^{x_2}$.

$$\begin{aligned}
K_o &:= R_1. \\
K_1 &:= K_o \vee (v_{x_2} = c_2) \\
&\equiv (v_{x_2} = a_2 \vee v_{x_2} = c_2). \\
K_2 &:= K_1 \vee (v_{x_2} = b_2) \\
&\equiv 1. \\
[R_1]^{x_2} &:= K_2.
\end{aligned}$$

Then $[R] := [R_1]^{x_2} \equiv 1$ as no new basic state trees can be added. The advantage of the new method is that the intermediate predicates are much simpler. By computing each branch of the state tree one by one to get some *intermediate fixpoints* such as $[R]^{x_1}$, we can control the size of BDD used to represent these intermediate predicates. In the computation of $[R]^{x_1}$, the atomic terms with state variable v_{x_2} remain unchanged. We find that this is the best way to apply the simplifying tautology as early as possible.

Now we can give the algorithm in the formal form.

Algorithm 4.2 Let $\mathbf{G} = (\mathbf{ST}, \mathcal{H}, \Sigma, \Delta, \mathbf{ST}_o, ST_m)$ be a state tree structure with $\mathbf{ST} = (X, x_o, \mathcal{T}, \mathcal{E})$. Let $R \in Pred(\mathbf{ST})$. Denote by $[R]^x$ the fixpoint for the superstate x. Then we have

```
function [R]^x :=
    K ← R
    if T(x) = or then
        repeat
            K' ← K
            K ← K ∨ (⋁_{σ_u ∈ Σ^x_{Iu}} Γ(K, σ_u))
            foreach y ∈ E(x) & T(y) ∈ {or, and}
```

```
8:              K ← [K]^y
9:          until K' ≡ K
10:     else if T(x) = and then
11:         repeat
12:             K' ← K
13:             foreach y ∈ E(x)
14:                 K ← [K]^y
15:         until K' ≡ K
16: end [R]^x
17: return K
```

◇

Remarks

1. $[R] := [R]^{x_o}$, the fixpoint of x_o.
2. The computation covers two cases. If x is an OR state, lines 4-9 compute the fixpoint. If x is an AND state, lines 11-15 do so. The computation will terminate as the change of K is monotone and the system is finite. It turns out that every algorithm in this chapter follows the same scheme.
3. The algorithm is recursive. The fixpoint $[R]^x$ depends on the computation of $[K]^y$, where K is an intermediate predicate and y is a superstate descendant of x, namely all of x's descendants have already been computed before computing $[R]^x$. The reason for sequencing the computations in this way is twofold. First, we want to apply the tautology

$$\left(\bigvee_{b \in \mathcal{E}(a)} (v_a = b)\right) \equiv 1$$

to any state variable v_a in the intermediate predicate K as early as possible, to control the size of intermediate predicates. If y is a superstate descendant of x, we want to compute $[K]^y$ first to try to quantify out the state variables v_y by applying the above tautology to v_y, and therefore make it possible also to apply the above tautology to v_x [9]. The tautology for every state variable is automatically applied by the BDD package, according to certain standard reduction rules. Second, by having a largest fixpoint $[K]^y$, we add into K as many basic state trees as possible from the computation of transitions under y. Then in the subsequent computation of another superstate z, each call to Γ can add more basic state trees to speed up the computation, because Γ is monotone in the sense that $K_1 \preceq K_2 \Rightarrow (\forall\sigma)\Gamma(K_1,\sigma) \preceq \Gamma(K_2,\sigma)$.

---

[9] Please refer to the definition of Θ where one can find that the tautology can be applied to the state variable v_x in the intermediate predicate K *only if* K is independent of all state variables under x.

4. To compute $[R]^x$, the algorithm requires iteration back and forth between x and its descendants. The fewer the number of iterations, the faster the algorithm terminates.

The structured Algorithm 4.2 has significant [10] performance improvement over that of the unstructured Algorithm 4.1. We will demonstrate this further in the next two chapters.

We can do better after some refinement of Algorithm 4.2. Here we introduce one improvement that is based on the observation of sharing events among AND components.

If the event sets of two generators are disjoint, we know that the synchronization of the two generators must result in a generator with the product number of states. On the other hand, if two generators have many shared events, intuition suggests that the synchronization of the two generators *may* result in a smaller generator. In summary, we may be better off considering two generators with disjoint event sets as *two independent units*, but two generators with many shared events as *one unit* during the synthesis process.

In a well-formed STS, events of the inner transition structure can only be shared among those holons matched to the children that are AND-adjacent to the same AND superstate (local coupling). So we just need to update lines 11-15 of Algorithm 4.2, the part for AND superstates. First we define the concept *cluster* as follows.

Definition 4.4 [cluster] Let $\mathbf{G} = (\mathbf{ST}, \mathcal{H}, \Sigma, \Delta, \mathbf{ST}_o, \mathcal{ST}_m)$ be a state tree structure with $\mathbf{ST} = (X, x_o, \mathcal{T}, \mathcal{E})$. Let x be an AND superstate on $\mathbf{G}$. Let $C := \{y | \mathcal{T}(y) = or \ \& \ x <_\times y\}$ be the set of OR states that are AND-adjacent to x. Then C can be partitioned by the family

$$\{C_i | i \in A \text{ some index set}\}$$

such that

1. every two holons matched to the superstates of different cells have disjoint internal event sets, i.e, $(\forall i, j \in A, i \neq j, x_i \in C_i, x_j \in C_j) \Sigma_I^{x_i} \cap \Sigma_I^{x_j} = \emptyset$, and
2. the partition is the finest according to event sharing, i.e., $(\forall i \in A, M \subset C_i)(\bigcup_{x_i \in M} \Sigma_I^{x_i}) \cap (\bigcup_{x_j \in (C_i - M)} \Sigma_I^{x_j}) \neq \emptyset$.

Call each cell C_i a *cluster* under x.

◇

Each cluster is an independent unit. Let $\mathbf{v}_{\mathbf{C_i}}$ be the set of variables assigned to all elements in C_i and their OR descendants. If we compute a fixpoint $[R]^{C_i}$ for cluster C_i, $[R]^{C_i}$ will have maximal number of basic state trees that can be added into $[R]^{C_i}$ from uncontrollable transitions inside and under all superstates of C_i. When we try to compute in another cluster C_j, we can add

[10] By significant we mean a hundred or even a thousand times faster.

more basic state trees into the resulting predicate at each Γ computation [11]. More importantly, the state variables in $\mathbf{v}_{\mathbf{C}_i}$ will stay unchanged during the computation in C_j (because the transitions in C_j cannot change the values of the variables in $\mathbf{v}_{\mathbf{C}_i}$ due to disjoint internal event sets) with the result that the size of intermediate predicates can be fairly controlled.

Now we can give our best available algorithm for computing $[\cdot]$.

Algorithm 4.3 Let $\mathbf{G} = (\mathbf{ST}, \mathcal{H}, \Sigma, \Delta, \mathbf{ST}_o, \mathcal{ST}_m)$ be a state tree structure with $\mathbf{ST} = (X, x_o, \mathcal{T}, \mathcal{E})$. Let $R \in Pred(\mathbf{ST})$. Denote by $[R]^x$ the fixpoint for the superstate x. Then:

1: **function** $[R]^x$:=
2: $\quad K \leftarrow R$
3: $\quad$ **if** $\mathcal{T}(x) = or$ **then**
4: $\quad\quad$ **repeat**
5: $\quad\quad\quad K' \leftarrow K$
6: $\quad\quad\quad K \leftarrow K \vee (\bigvee_{\sigma_u \in \Sigma^x_{Iu}} \Gamma(K, \sigma_u))$
7: $\quad\quad\quad$ **foreach** $y \in \mathcal{E}(x)$ & $\mathcal{T}(y) \in \{or, and\}$
8: $\quad\quad\quad\quad K \leftarrow [K]^y$
9: $\quad\quad$ **until** $K' \equiv K$
10: $\quad$ **else if** $\mathcal{T}(x) = and$ **then**
11: $\quad\quad$ **repeat**
12: $\quad\quad\quad K' \leftarrow K$
13: $\quad\quad\quad$ **foreach cluster** C_i **under** x
14: $\quad\quad\quad\quad$ **repeat**
15: $\quad\quad\quad\quad\quad K'' \leftarrow K$
16: $\quad\quad\quad\quad\quad$ **foreach** $y \in C_i$
17: $\quad\quad\quad\quad\quad\quad K \leftarrow [K]^y$
18: $\quad\quad\quad\quad$ **until** $K'' \equiv K$
19: $\quad\quad$ **until** $K' \equiv K$
20: **end** $[R]^x$
21: **return** K

$\diamond$

The only difference between Algorithm 4.2 and 4.3 is that the latter adds lines 14-18 to compute a largest fixpoint for each cluster. However, the performance improvement is significant. By use of Algorithm 4.3 in our experiment on the AIP example, the size of intermediate predicates was around 10 times smaller and the computing time was around 15 times faster.

---

[11] Notice that Γ is monotone in the sense that $P_1 \preceq P_2 \Rightarrow (\forall\sigma)\Gamma(P_1, \sigma) \preceq \Gamma(P_2, \sigma)$.

4.3.3 Computation of $CR(\mathbf{G}, \cdot)$

From the definition of $CR(\mathbf{G}, \cdot)$, we can compute it from the basic algorithm given as follows.

Algorithm 4.4 [Basic algorithm for $CR(\mathbf{G}, \cdot)$] Let $\mathbf{G} = (\mathbf{ST}, \mathcal{H}, \Sigma, \Delta, P_o, P_m)$ be a state tree structure.
$CR(\mathbf{G}, P)$

1. $K_o := P \wedge P_m$.
2. $K_{i+1} := K_i \vee (P \wedge \bigvee_{\sigma \in \Sigma} \Gamma(K_i, \sigma))$.
3. If $K_{n+1} \equiv K_n$, then $CR(\mathbf{G}, P) := K_n$. Otherwise go back to step 2.

◇

Algorithm 4.4 is similar to Algorithm 4.1. So we can follow the same reasoning to get the structured Algorithm 4.5.

Algorithm 4.5 Let $\mathbf{G} = (\mathbf{ST}, \mathcal{H}, \Sigma, \Delta, P_o, P_m)$ be a state tree structure. Let $P, R \in Pred(\mathbf{ST})$. Denote by $CR^x(\mathbf{G}, P, R)$ the fixpoint for the superstate x such that for each $b_o \models CR^x(\mathbf{G}, P, R)$, there is a sequence $\{b_i | i = 0, 1, \ldots, n\}$ for b_o to reach $b_n \models R$ by local transitions in x and every b_i satisfies P. Precisely, we have:

function $CR^x(\mathbf{G}, P, R) :=$
        $K \leftarrow R$
    if $\mathcal{T}(x) = or$ **then**
        repeat
            $K' \leftarrow K$
            $K \leftarrow K \vee (P \wedge \bigvee_{\sigma \in \Sigma_I^x} \Gamma(K, \sigma))$
            foreach $y \in \mathcal{E}(x)$ & $\mathcal{T}(y) \in \{or, and\}$
                $K \leftarrow CR^y(\mathbf{G}, P, K)$
        until $K' \equiv K$
    else if $\mathcal{T}(x) = and$ **then**
        repeat
            $K' \leftarrow K$
            foreach cluster C_i **under** x
                repeat
                    $K'' \leftarrow K$
                    foreach $y \in C_i$
                        $K \leftarrow CR^y(\mathbf{G}, P, K)$
                until $K'' \equiv K$
        until $K' \equiv K$
end $CR^x(\mathbf{G}, P, K)$
return K

◇

Remarks

1. $CR(\mathbf{G}, P) := CR^{x_o}(\mathbf{G}, P, P \wedge P_m)$.
2. For convenience, call the fixpoint computed by lines 14-18 $CR^{C_i}(\mathbf{G}, P, K)$.
3. This is a recursive algorithm. It requires iteration back and forth between x and its descendants. The smaller the number of iterations, the faster the algorithm terminates.
4. The structure of the Algorithm is same as that of Algorithm 4.3.

Inference

From our observation, the computation of $CR(\mathbf{G}, P)$ is usually more difficult than $[P]$ as

1. we just need to compute Γ for those uncontrollable events for $[P]$, and the size of Σ_u is smaller than Σ;
2. we have to guarantee $CR(\mathbf{G}, P) \preceq P$ by performing the additional meet operation of P and the resulting predicates from Γ at each step.

In the synthesis of complex systems, most computing time is spent on computing $CR(\mathbf{G}, P)$. So in this subsection, we explore another option called *inference* to speed up the computation of $CR(\mathbf{G}, P)$.

Checking the structured Algorithm 4.5, to get $CR^x(\mathbf{G}, P, R)$ we need to compute the fixpoints for all of x's superstate descendants. But it would be convenient and economical if we could *infer* from some *knowledge* of $\mathbf{G}$ and P, the fixpoints of those children instead of computing them.

Figure 4.21 illustrates the idea. In the picture, P and A are predicates that can be identified by two sets of states. x is a state that satisfies A. Because the computation of $CR(\mathbf{G}, P)$ is backward, the first state of A encountered by the computation is x. After x is added into $CR(\mathbf{G}, P)$, usually we need to

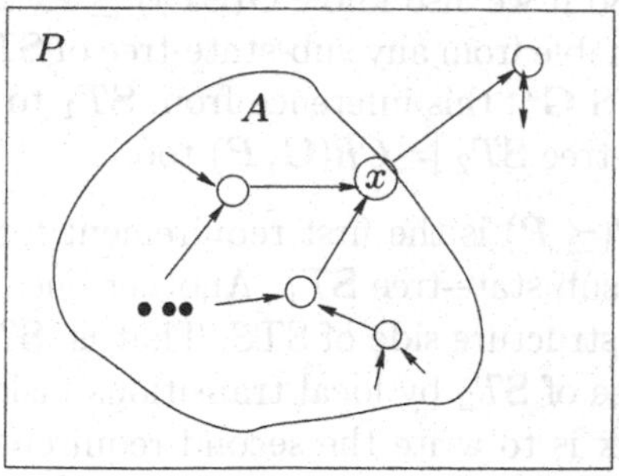

Fig. 4.21. Conditions for the inference to be valid

call Γ enough times to add the remaining states in A to $CR(\mathbf{G}, P)$. However, the approach in this section is to investigate the structure of A first. If A has some nice properties that can imply $A \preceq CR(\mathbf{G}, P)$ from the fact that $x \models CR(\mathbf{G}, P)$, we no longer need to call Γ to compute each state inside A. This will of course speed up the computation.

Formally, a sufficient condition for the inference

$$x \models CR(\mathbf{G}, P) \Rightarrow A \preceq CR(\mathbf{G}, P)$$

is given directly from the definition of $CR(\mathbf{G}, P)$:

1. $A \preceq P$. This is a necessary condition for $A \preceq CR(\mathbf{G}, P)$. Also
2. x is *locally reachable* from any other states in A. x is locally reachable from y if there exists a path from y to x *and* all states on the path belong to A (and therefore satisfy P).

This inference is useful because it deduces the coreachability of the entire set A, the global property of A, from the coreachability of a *critical* state x in A and the local reachability to x.

However, the challenging part of this story is to write the above condition in a predicate! Fortunately, we can do this neatly in our STS setting. It is necessary that the predicate should include the information from P and the STS $\mathbf{G}$.

First, let's look at the predicate P and try to get the predicate A. For a superstate x, denote by $\mathbf{v_x}$ the set of all state variables assigned to OR superstates in $\mathbf{ST}^x$. Then in P,

$$A := (\forall \mathbf{v_x})P$$

is the weakest predicate that enjoys the following useful property. [12]

- If $\mathbf{ST}_1 \models (\forall \mathbf{v_x})P$, then $\mathbf{ST}_2 \models P$, where $\mathbf{ST}_2$ is obtained by replacing $\mathbf{ST}_1^x$ with $\mathbf{ST}^x$ (adding all missing branches under x to $\mathbf{ST}_1$ to get $\mathbf{ST}_2$). Notice that $\mathbf{ST}_1 \leq \mathbf{ST}_2$. So if we also know $\Theta(\mathbf{ST}_2) \preceq P$, $\mathbf{ST}_1 \models CR(\mathbf{G}, P)$ and $\mathbf{ST}_1$ is locally reachable from any sub-state-tree of $\mathbf{ST}_2$ by local transitions inside the child STS $\mathbf{G}^x$, this inference from $\mathbf{ST}_1$ to $\mathbf{ST}_2$ tells us that the "bigger" sub-state-tree $\mathbf{ST}_2 \models CR(\mathbf{G}, P)$ too.

The predicate $(\forall \mathbf{v_x})P (\preceq P)$ is the first requirement for such an inference to be carried out on the sub-state-tree $\mathbf{ST}_1$. Another one that $\mathbf{ST}_1$ must satisfy is from the transition structure side of STS. That is, $\mathbf{ST}_1$ is locally reachable from any sub-state-tree of $\mathbf{ST}_2$ by local transitions inside the child STS $\mathbf{G}^x$.

The remaining task is to write the second requirement by predicate, i.e., to find the critical sub-state-trees. After that, the inference can be carried out by a neat logical formula. Based on the structure of the recursive algorithms we had, there are two cases that we need to take care of. Notice that the

[12] One may define A as any subpredicate of P. But to us this is the most obvious one having the useful property.

inference on the root state is useless because there the predicate P must be 1, logical truth, in order for the inference to be valid. So we only consider those AND/OR superstates that have ancestors.

1. x is an OR superstate and a child of another OR superstate.
The internal event set of H^x is disjoint with that of any other holons. As we know, a sub-state-tree moves out from H^x if it has a state $p \in X_m^x$ and an outgoing transition of H^x is triggered. Then in general [13] it will be one step "closer" to those marker state trees. Therefore, the *critical sub-state-tree* $\mathbf{ST}_{c,x}$ for the superstate x should have those states in X_m^x, the states at the interface of H^x. [14] Formally, $a \in \mathbf{ST}_{c,x}$ if and only if
a) $a|x$. a is parallel with x.
b) $a \leq x$.
c) $x \leq a \Leftrightarrow (a \in X_m^x \ \& \ (CR^x(\mathbf{G}, true, v_x = a) = 1))$.
From the Algorithm 4.5, $(CR^x(\mathbf{G}, true, v_x = a) = 1)$ if and only if a is locally reachable in the child STS $\mathbf{G}^x$. $\mathbf{ST}_{c,x} = \emptyset$, the empty state tree if no descendants of x are on $\mathbf{ST}_{c,x}$, i.e., no states are found in (c). Then $\Theta(\mathbf{ST}_{c,x})$ is the predicate for the second requirement.
Combining the two predicates $(\forall \mathbf{v_x})P$ and $\Theta(\mathbf{ST}_{c,x})$, we can define the *generated transition relation of x under the guard of P* [15] by

$$N^x(P) := \Theta(\mathbf{ST}_{c,x}) \wedge (\forall \mathbf{v_x})P,$$

where as usual $\mathbf{v_x}$ is the set of variables assigned to those OR superstates in $\mathbf{ST}^x$. For fixed state x, $N^x(P)$ is a predicate transformer.
A tree satisfying $N^x(P)$ will be a sub-state-tree of $\mathbf{ST}_{c,x}$ and also satisfy $(\forall \mathbf{v_x})P$. Thus $N^x(P)$ is the condition for the inference we mentioned before to be carried out at the superstate x.
Let K be the intermediate predicate in the computation of $CR^{x_0}(\mathbf{G}, P, P \wedge P_m)$. The inference result at x is given simply by

$$\exists v_x (K \wedge N^x(P)).$$

No other variables in $\mathbf{v_x}$ will appear in $K \wedge N^x(P)$ because for all descendants of x, only some *simple state children of x* are allowed to be on $\mathbf{ST}_{c,x}$. Thus we just need $\exists v_x$ instead of $\exists \mathbf{v_x}$ to quantify out all variables of $\mathbf{ST}^x$.
Notice that

$$\exists \mathbf{v_x} K$$

[13] Not always because sometimes it moves away from marker state trees if there are lower level states on the marker state trees.

[14] One may include any possible states in X^x instead of just in X_m^x. But we think that is not as useful because the inference usually first occurs on the boundary states during the backward computation of $CR(\mathbf{G}, P)$.

[15] We call it a *transition relation* just like N_σ for a given event σ. The reason is that the formula of the inference given later is similar to that of Γ.

is the weakest predicate that we can get from computing $CR^x(\mathbf{G}, P, K)$.[16] So the condition that we can avoid calling $CR^x(\mathbf{G}, P, K)$ to compute inside x is when

$$\exists v_x(K \wedge N^x(P)) \equiv \exists \mathbf{v_x} K,$$

i.e., we have obtained the largest fixpoint for x by inference. So if the above equality holds, we know in advance that

$$CR^x(\mathbf{G}, P, K) := \exists v_x(K \wedge N^x(P))$$

without going into x.

2. x is an AND superstate and a child of another OR superstate. Then any OR state that is AND-adjacent to x must be in a cluster that has a unique set of internal events.
   The objective is to find
   a) the inference condition to avoid calling $CR^x(\mathbf{G}, P, K)$ (the ideal case) and
   b) the inference condition to avoid calling $CR^{C_m}(\mathbf{G}, P, K)$ for each cluster C_m.

   Just as in the first case when x is an OR superstate, only the outgoing boundary states are interesting to us. Owing to boundary consistency, all holons matched to the OR states that are AND-adjacent to x must have exactly the same outgoing transitions. Because the system must be deterministic, each outgoing transition of a child holon must be labelled by a unique event. So it is easy to see that all such holons must share the same outgoing event set Σ^x_{BO}. For each event $\sigma \in \Sigma^x_{BO}$, denote the *critical sub-state-tree of σ for the superstate x* by $\mathbf{ST}_{\sigma,x}$. Let $C := \{y_1, y_2, \ldots, y_n\}$ be the set of OR states that are AND-adjacent to x. Then we have $a_i \in \mathbf{ST}_{\sigma,x}$ if and only if
   a) $a_i | x$. a_i is parallel with x.
   b) $a_i \leq x$.
   c) $(\exists y \in C) x < a_i \leq y$.
   d) $y_i \leq a_i$ implies $\delta^{y_i}(a_i, \sigma)!$ & $\forall (j \neq i) \exists (a_j \in X^{y_j}_m)(\delta^{y_j}(a_j, \sigma)!$ & $(CR^x(\mathbf{G}, 1, \bigwedge_{\forall k} v_{y_k} = a_k) \equiv 1))$.

   $\mathbf{ST}_{\sigma,x} = \emptyset$, the empty state tree if no states are found in (d). Then the *generated transition relation of x under the guard of P* is given by .

$$N^x(P) := (\bigvee_{\sigma \in \Sigma^x_{BO}} \Theta(\mathbf{ST}_{\sigma,x})) \wedge (\forall \mathbf{v_x}) P,$$

   where as usual $\mathbf{v_x}$ is the set of variables assigned to those OR superstates in ST^x.

---

[16] Think of K as a set of sub-state-trees. As the events of $\mathbf{G}^x$ are local and therefore will not change any other branches of each tree in K except the branches under superstate x, the weakest predicate we can get is by replacing the child state tree rooted by x of each element in K with the child state tree $\mathbf{ST}^x$ (by adding all missing branches under x). $\exists \mathbf{v_x} K$ does that job.

Applying the same approach to each cluster C_m under x, we can define the *critical sub-state-tree of* σ *for the cluster* C_m $\mathbf{ST}_{\sigma,C_m}$ by
$a_i \in \mathbf{ST}_{\sigma,C_m}$ if and only if
a) $a_i|C_m$. a_i is parallel with any elements in C_m.
b) $a_i \leq x$.
c) $(\exists y \in C_m)a_i \leq y$.
d) $(\exists y_i \in C_m)y_i \leq a_i$ implies

$$\delta^{y_i}(a_i,\sigma)! \ \& \ \forall(y_j \in C_m - \{y_i\})\exists(a_j \in X_m^{y_j})$$
$$(\delta^{y_j}(a_j,\sigma)! \ \& \ (CR^{C_m}(\mathbf{G},1, \bigwedge_{\forall y_k \in C_m} v_{y_k} = a_k) \equiv 1)).$$

$\mathbf{ST}_{\sigma,C_m} = \emptyset$, the empty state tree if no states are found in (d). Then the *generated transition relation of* C_m *under the guard of* P by

$$N^{C_m}(P) := (\bigvee_{\sigma \in \Sigma_{BO}^x} \Theta(\mathbf{ST}_{\sigma,C_m})) \wedge (\forall \mathbf{v}_{\mathbf{C_m}})P,$$

where $\mathbf{v}_{\mathbf{C_m}}$ is the set of variables assigned to those OR superstates in C_m and their OR descendants.
Now we know in advance that

$$CR^x(\mathbf{G},P,K) := \exists v_x(K \wedge N^x(P))$$

if

$$\exists v_x(K \wedge N^x(P)) = \exists \mathbf{v_x} K$$

and for each cluster C_m of x,

$$CR^{C_m}(\mathbf{G},P,K) := \exists v_{C_m}(K \wedge N^{C_m}(P))$$

if

$$\exists v_{C_m}(K \wedge N^{C_m}(P)) \equiv \exists \mathbf{v}_{\mathbf{C_m}} K,$$

where v_{C_m} is the set of variables only assigned to those OR superstates in C_m.
Because the inference condition of x is dependent on that of its clusters, we can write a program to obtain their predicate formulas together to avoid repeating computation.

Applying the inference, we give the updated algorithm as follows.

Algorithm 4.6 Let $\mathbf{G} = (\mathbf{ST}, \mathcal{H}, \Sigma, \Delta, P_o, P_m)$ be a state tree structure. Let $P, R \in Pred(\mathbf{ST})$. Denote by $CR^x(\mathbf{G},P,R)$ the fixpoint for the superstate x such that for each $b_o \models CR^x(\mathbf{G},P,R)$, there is a sequence $\{b_i|i = 0,1,\ldots,n\}$ for b_o to reach $b_n \models R$ by local transitions in x and every b_i satisfies P. Precisely, we have

```
function CR^x(G, P, R) :=
    K ← R
    if T(x) = or then
        repeat
            K' ← K
            K ← K ∨ (P ∧ ⋁_{σ∈Σ_I^x} Γ(K, σ))
            foreach y ∈ E(x) & T(y) ∈ {or, and}
                if ∃v_y(K ∧ N^y(P)) ≡ ∃v_y K
                    K ← ∃v_y(K ∧ N^y(P))
                else
                    K ← CR^y(G, P, K)
        until K' ≡ K
    else if T(x) = and then
        repeat
            K' ← K
            foreach cluster C_i under x
                if ∃v_{C_i}(K ∧ N^{C_i}(P)) ≡ ∃v_{C_i} K
                    K ← ∃v_{C_i}(K ∧ N^{C_i}(P))
                else
                    K ← CR^{C_i}(G, P, K)
        until K' ≡ K
end CR^x(G, P, K)
return K
```

$\diamond$

4.4 Controller Implementation

In the original RW framework, a controller can be implemented by an automaton. The controller performs two tasks. The first is to keep track of state changes occurring in the plant; the second is to make control decisions based on the current plant state. The control diagram is illustrated in Figure 4.22,

This is a very nice theoretical implementation because both plant and controller can be represented as automata. However, as to complex systems, especially locally coupled systems (with fewer shared events), the price of the first task, recording state change, can be very high. The centralized controller for a moderately complex system can easily have millions of states.

We trace the problem to the fact that a centralized controller in the RW framework has to do two tasks by one agent(automaton). One natural conclusion is to have two agents in the controller. One records the plant's state changes, the other issues control decisions. The new control diagram is given in Figure 4.23. Agent 1 (tracker) tells agent 2 (decsion maker) which state the plant is in. Agent 2 does disablement based on this information.

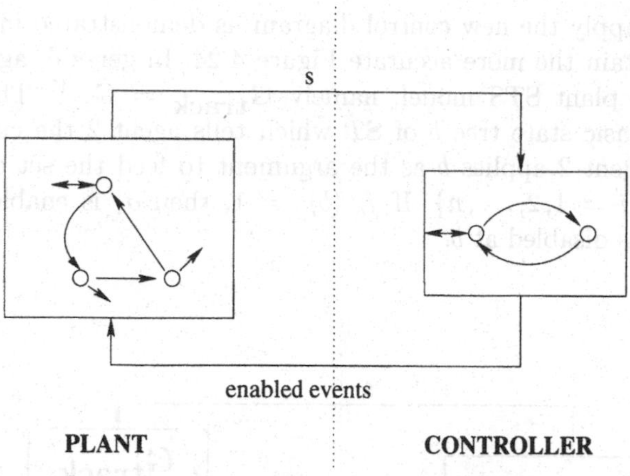

Fig. 4.22. Original Control Diagram in the RW Framework

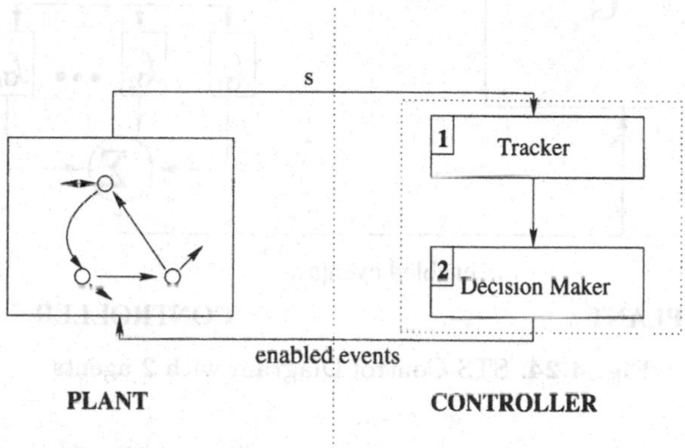

Fig. 4.23. Control Diagram with 2 agents

Now we supply further details on applying the above two-agents control diagram to our STS control.

Given a control problem $(\mathbf{G}, P)$ with $\mathbf{G} = (\mathbf{ST}, \mathcal{H}, \Sigma, \Delta, P_o, P_m)$, as shown in the previous section, we compute the optimal controlled behavior as a predicate, defined by

$$C := \sup \mathcal{C}^2\mathcal{P}(\neg P),$$

where P is the predicate for the set of *illegal* sub-state-trees. If the closed system under control is nonempty, i.e.,

$$P_o \wedge C \not\equiv 0,$$

the control itself can be easily implemented based on the following set of predicates $\{f_\sigma\}$, defined by

$$(\forall \sigma \in \Sigma_c) f_\sigma(b) := \begin{cases} 0, \text{ if } \Delta(b, \sigma) \models \neg C \\ 1, \text{ otherwise} \end{cases} .$$

Then we can apply the new control diagram as demonstrated in Figure 4.23 to STS, to obtain the more accurate Figure 4.24. In general, agent 1 is just a copy of the plant STS model, namely $\mathbf{G_{track}} = \mathbf{G}$.[17] The output of agent 1 is a basic state tree b of $\mathbf{ST}$ which tells agent 2 the current status of $\mathbf{G}$. Then agent 2 applies b as the argument to feed the set of predicates $\{f_{\sigma_i} | \sigma_i \in \Sigma_c, i = 1, 2, \ldots, n\}$. If $f_{\sigma_i}(b) := 1$, then σ_i is enabled to occur. Otherwise σ_i is disabled at b.

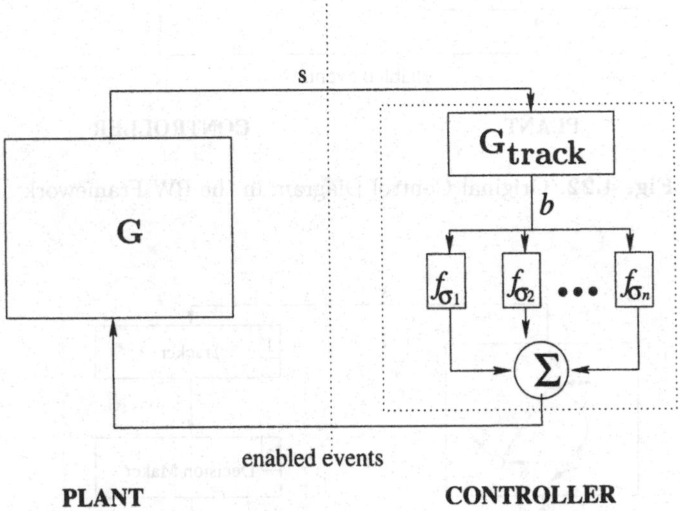

Fig. 4.24. STS Control Diagram with 2 agents

Comparing it with the original diagram in the RW framework, we see that the reactive speed of our controller will depend on how fast $f_\sigma(b)$ can be computed. This is important in real-time control. Notice that to compute $f_\sigma(b)$, it is required to call $\Delta(b, \sigma)$ and then verify the satisfiability of $\Delta(b, \sigma) \models \neg C$. This may be too slow for real-time control. So we decide to compute $\{f_\sigma\}$ off-line.

For any controllable event σ, we know $\Delta(\text{Elig}_{\mathbf{G}}(\sigma), \sigma) = \text{Next}_{\mathbf{G}}(\sigma)$, the largest sub-state-tree of $\mathbf{ST}$ that is targeted by σ. Given the optimal controlled behavior C, the predicate $\Theta(\text{Next}_{\mathbf{G}}(\sigma))$ is divided into two subpredicates,

$$N_{good} := \Theta(\text{Next}_{\mathbf{G}}(\sigma)) \wedge C,$$

the legal subpredicate of $\Theta(\text{Next}_{\mathbf{G}}(\sigma))$ and

$$N_{bad} := \Theta(\text{Next}_{\mathbf{G}}(\sigma)) \wedge \neg C,$$

the illegal subpredicate of $\Theta(\text{Next}_{\mathbf{G}}(\sigma))$. This is illustrated in Figure 4.25.

[17] $\mathbf{G_{track}}$ can also be a simplified version of the plant STS model $\mathbf{G}$. For example, if all control functions $\{f_{\sigma_i}\}$ do not depend on some state variables, then agent 1 does not need to track their representing holons.

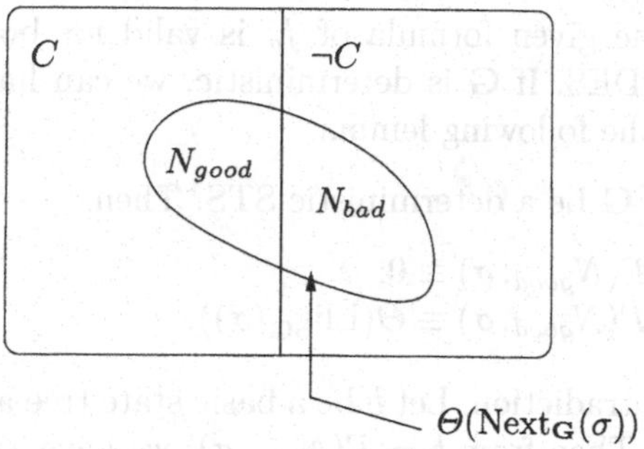

Fig. 4.25. Good predicate N_{good} and bad predicate N_{bad}: $\Theta(\text{Next}_{\mathbf{G}}(\sigma)) := N_{good} \vee N_{bad}$

Suppose that at a certain basic state tree b, the controller must disable σ if $\Delta(b, \sigma) \models N_{bad}$. So starting from N_{bad}, we know that the predicate

$$\Gamma(N_{bad}, \sigma)$$

is identified by the largest set of basic state trees where σ must be disabled. Then it is easy to come up with an explicit formula for f_σ:

$$f_\sigma := (\neg\Gamma(N_{bad}, \sigma)) \wedge \Theta(\text{Elig}_{\mathbf{G}}(\sigma)).$$

The first part $\neg\Gamma(N_{bad}, \sigma)$ includes all basic state trees that are not led to illegal trees in N_{bad} by transitions labelled with σ, from the negation operator. The second part $\Theta(\text{Elig}_{\mathbf{G}}(\sigma))$ makes the description more accurate, i.e., the event σ must be eligible too. For many controllable events, we can omit the second part in the above formula to get a simpler predicate. But one must be very careful. For example, the second part of f_{α_2} is necessary for the Small Factory in Figure 4.3 of page 80. Notice that the final synthesis is based on the plant model that already disables α_2 when the buffer is empty. This must be enforced in f_{α_2} by meeting with $\Theta(\text{Elig}_{\mathbf{G}}(\alpha_2))$, where $\mathbf{G}$ is the STS in Figure 4.3.

To illustrate the correctness of our formula, we apply it to two special cases.

1. $N_{bad} := 0$, i.e., all target trees of σ, if exist, are legal under C. We have

$$\begin{aligned} f_\sigma &:= \neg\Gamma(0, \sigma) \wedge \Theta(\text{Elig}_{\mathbf{G}}(\sigma)) \\ &:= \Theta(\text{Elig}_{\mathbf{G}}(\sigma)) \end{aligned}$$

as required.

2. $N_{bad} := 1$, i.e., all target trees of σ are illegal under C. We have

$$\begin{aligned} f_\sigma &:= \neg\Gamma(1, \sigma) \wedge \Theta(\text{Elig}_{\mathbf{G}}(\sigma)) \\ &:= \neg\Theta(\text{Elig}_{\mathbf{G}}(\sigma)) \wedge \Theta(\text{Elig}_{\mathbf{G}}(\sigma)) \\ &:= 0 \end{aligned}$$

as required.

Notice that the given formula of f_σ is valid for both deterministic and nondeterministic DES. If $\mathbf{G}$ is deterministic, we can have a simpler formula for f_σ, based on the following lemma.

Lemma 4.4 Let $\mathbf{G}$ be a deterministic STS. Then

1. $\Gamma(N_{bad}, \sigma) \wedge \Gamma(N_{good}, \sigma) \equiv 0$;
2. $\Gamma(N_{bad}, \sigma) \vee \Gamma(N_{good}, \sigma) \equiv \Theta(\text{Elig}_{\mathbf{G}}(\sigma))$.

Proof. 1. By contradiction. Let b be a basic state tree and $b \models (\Gamma(N_{bad}, \sigma) \wedge \Gamma(N_{good}, \sigma))$. Then from $b \models \Gamma(N_{bad}, \sigma)$, we have $\Delta(b, \sigma)$ is a basic state tree and $\Delta(b, \sigma) \models N_{bad}$. Also from $b \models \Gamma(N_{good}, \sigma)$, we have $\Delta(b, \sigma) \models N_{good}$. Thus, $\Delta(b, \sigma) \models (N_{good} \wedge N_{bad})$. This contradicts the fact that $N_{good} \wedge N_{bad} \equiv 0$.
2. Automatically $\Gamma(N_{bad}, \sigma) \vee \Gamma(N_{good}, \sigma) \preceq \Theta(\text{Elig}_{\mathbf{G}}(\sigma))$. Now prove "$\succeq$". Let b be a basic state tree and $b \models \Theta(\text{Elig}_{\mathbf{G}}(\sigma))$. Then $\Delta(b, \sigma)$ is a basic state tree and $\Delta(b, \sigma) \models N_{good} \vee N_{bad}$ because $N_{good} \vee N_{bad} = \Theta(\text{Next}_{\mathbf{G}}(\sigma))$. If $\Delta(b, \sigma) \models N_{good}$, then $b \models \Gamma(N_{good}, \sigma)$. Otherwise $b \models \Gamma(N_{bad}, \sigma)$. Thus $b \models \Gamma(N_{good}, \sigma) \vee \Gamma(N_{bad}, \sigma)$ as required.

Now we give the new formula as

$$f_\sigma := \Gamma(N_{good}, \sigma).$$

Comparing it with the original one, we find that this new formula enables σ at a smaller number of trees, because $\Gamma(N_{good}, \sigma) \preceq \neg\Gamma(N_{bad}, \sigma)$ from item 1 of the above lemma. But all basic state trees in $(\neg\Gamma(N_{bad}, \sigma) - \Gamma(N_{good}, \sigma))$ are in fact don't-care trees where σ is not eligible, because of item 2 of the above lemma. So this formula is correct for deterministic DES. The new formula *only enables* at those trees that it has to, whereas the old formula *only disables* at those trees that it has to. Both will ensure the correct controlled behavior.

The benefit we enjoy is a simpler formula and less computation, i.e., no negation operations in computing N_{good} and f_σ, and no intersection with $\Theta(\text{Elig}_{\mathbf{G}}(\sigma))$ because $\Gamma(N_{good}, \sigma) \preceq \Theta(\text{Elig}_{\mathbf{G}}(\sigma))$.

A BDD simplification tool is available in our program to further simplify the control logic of f_σ. There is an elegant procedure called *bdd_simplify* in our BDD package that, given two BDD u, d, computes a smaller BDD u' such that $u \wedge d \equiv u' \wedge d$. The computational complexity is $O(|u| \times |d|)$. Notice that f_σ is meaningfully applied at the basic state tree b only when $b \models C \wedge \Theta(\text{Elig}_{\mathbf{G}}(\sigma))$, where C is the BDD for the set of all legal basic state trees. So we can let the domain BDD $d := C \wedge \Theta(\text{Elig}_{\mathbf{G}}(\sigma))$ and apply the above procedure to get a control function f'_σ, with fewer BDD nodes, such that $f'_\sigma \wedge d \equiv f_\sigma \wedge d$.

We will apply the new formula to all examples given in this book because all of our STS are deterministic. After computing f_σ off-line, the reaction time of our controller can be very fast as the verification time of $b \models f_\sigma$ is linear in the number of BDD nodes of f_σ.

4.5 Tutorial Examples

In this section, we give two simple control problems and present our resulting controllers, represented by a set of control functions.

4.5.1 The Small Factory

The STS model of the Small Factory is given in Figure 4.26. We only take care of the buffer overflow and underflow here. To fulfil this, we need to disable event 3 at state 0 of the holon BUF and disallow event 2 to occur at state 1 of the holon BUF. The event 2 is uncontrollable, which leads to the illegal predicate

$$v_{\mathrm{M1}} = 1 \wedge v_{\mathrm{BUF}} = 1.$$

The optimal controlled behavior is given in Figure 4.27. The dotted line is with value 0 and the solid line with value 1. So the figure shows that the only illegal RW states are the ones with both M1 and BUF at state 1 (there are two such basic state trees).

The control function f_3 for event 3, shown in Figure 4.28 is the same as $\Theta(\mathrm{Elig}_{\mathrm{SF}}(3))$, i.e., it is enabled whenever it is eligible in SF. We need the term $\Theta(\mathrm{Elig}_{\mathrm{SF}}(3))$ because the event 3 is already disabled in the STS model SF before synthesis.

The control function f_1 for event 1, shown in Figure 4.29, is a simplified one (by using our *bdd_simplify* procedure), because we do not need the intersection with $\Theta(\mathrm{Elig}_{SF}(1))$. It is remarkable because the BDD graph tells us the simple fact that event 1 is enabled only when the buffer is empty, i.e., at state 0

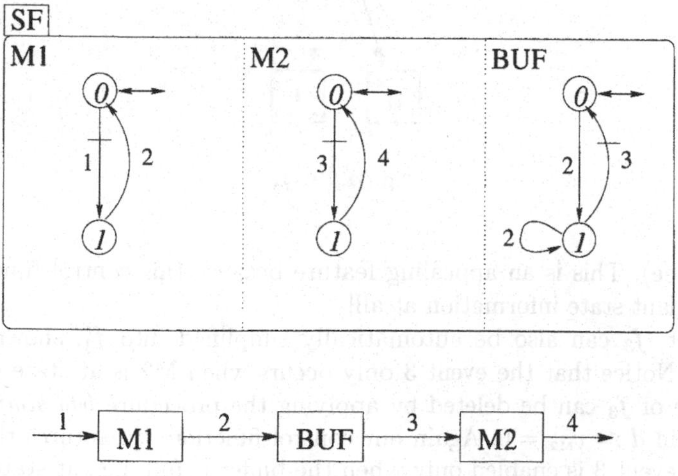

Fig. 4.26. STS model of the Small Factory

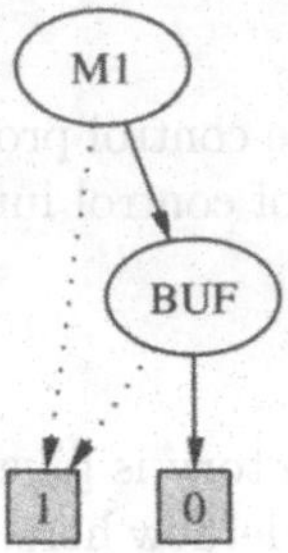

Fig. 4.27. Optimal controlled behavior BDD of the Small Factory

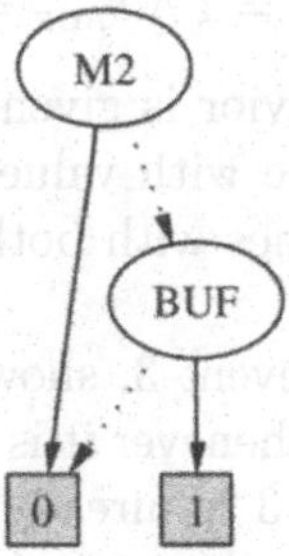

Fig. 4.28. f_3

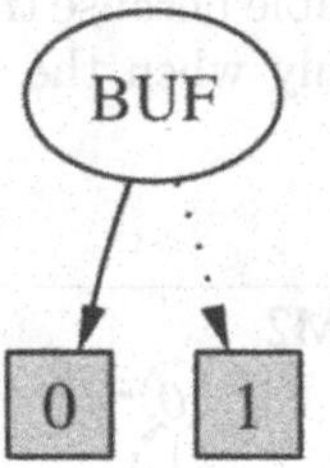

Fig. 4.29. f_1

(dotted line). This is an appealing feature because this control function has *no* redundant state information at all!

In fact, f_3 can also be automatically simplified into f_3', shown in Figure 4.30. Notice that the event 3 only occurs when M2 is at state 0. So the root node of f_3 can be deleted by applying the procedure *bdd_simplify* with the domain $d := v_{\mathrm{M2}} = 0$. Again our control function f_3' captures the simple fact that event 3 is enabled only when the buffer is full, i.e., at state 1 (solid line).

Finally, we can claim that we have found the best control functions for the Small Factory.

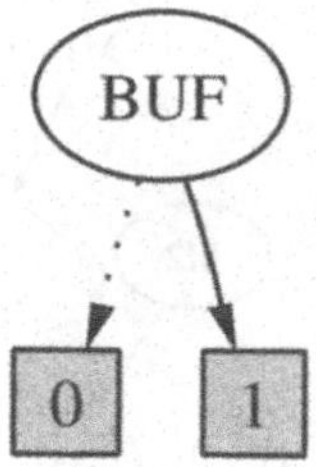

Fig. 4.30. f_3'

4.5.2 The Transfer Line

The STS model of the Transfer Line example (chapter 4 of [Won04]) is given in Figure 4.31.

The optimal controlled behavior is given in Figure 4.32. It shows that the legal RW states are the ones with at most one AND component that can stay at state 1.

There are 3 controllable events, 1, 3 and 5. We draw their original control functions computed by the new formula $\Gamma(N_{good}, \sigma)$ in Figure 4.33.

Then we apply the *bdd_simplify* procedure to compute the simplified control functions, given in Figure 4.34.

This is the simplest version we can ever get! Event 3 is enabled if and only if the buffer 1 is full. Event 5 is enabled if and only if the buffer 2 is full. Event 1 is enabled if and only if no other workpiece is still in the system.

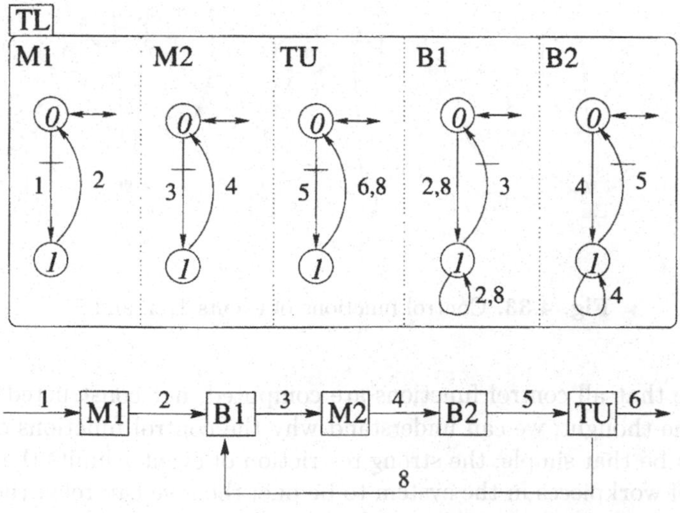

Fig. 4.31. STS model of the Transfer Line

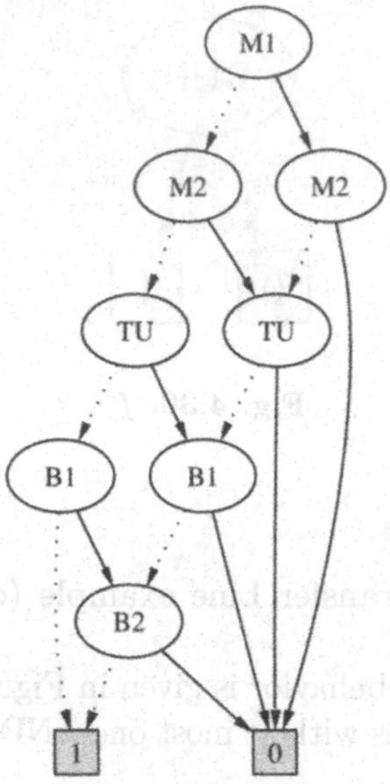

Fig. 4.32. Optimal controlled behavior BDD of the Transfer Line

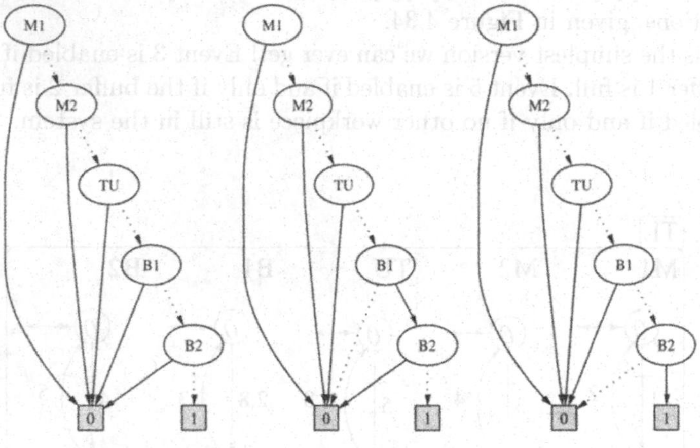

Fig. 4.33. Control functions of events 1, 3, and 5

Notice that all control functions are computed, not constructed by hand. With some thought, we can understand why the control functions of event 3 and 5 can be that simple: the strong restriction of event 1 limits the maximal number of workpieces in the system to be one; then we can relax the disablement of events 3 and 5 accordingly. This decision made by our program could be considered 'intelligent'.

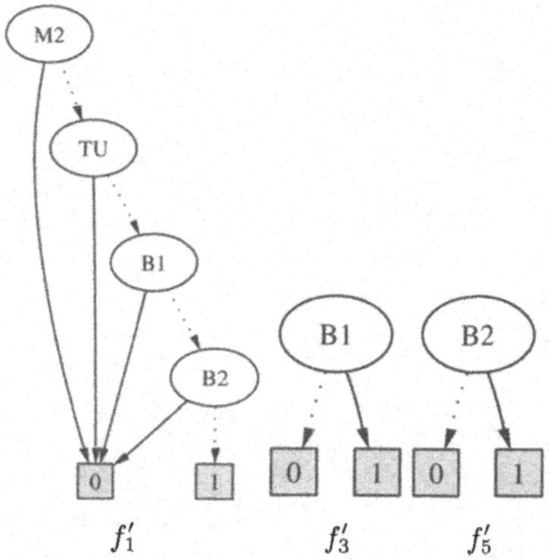

Fig. 4.34. Simplified Control functions of events 1, 3, and 5

4.6 Summary

This chapter is the core of the book. We have demonstrated the symbolic representation and synthesis of STS control problems.

By introducing the recursive function Θ, we can encode the hierarchical state space and transition structures easily. The recursive symbolic synthesis benefits from the rich structure of the STS model. Some techniques, e.g., clusters and inference, are also introduced to make the computation even more efficient. The control implementation in our framework is (we think) elegant. We profitably exploit the computational power of BDD throughout the whole chapter.

Additional techniques for logic formula manipulation can be developed in our framework to further speed up the computation. Also notice that the resulting controller C is a centralized one. A decentralized synthesis strategy may be a better path to attack even larger complex systems. But all in all, symbolic computation will provide researchers with a sound foundation for moving ahead.

5

The Production Cell Example

This chapter explains how to model and control a benchmark example, the Production Cell, using the State Tree Structure (STS) methodology. We will focus on describing a complex control problem in our framework, i.e., how to model a complex system top-down and write accurate logic specification formulas for it.

5.1 Introduction

The production cell benchmark is of interest to formal method researchers as it is complicated but still manageable. Many results, mostly in the model checking area, are presented in their official website [1]. But no supervisory control has been given. The main reason is the complexity in both the modelling and synthesis stages.

The production cell, shown in Figure 5.1, consists of six interconnected parts: feed belt, elevating rotary table, robot, press, deposit belt and travelling crane. One noticeable feature is that the robot has two arms to maximize the capacity of the press, namely to make it possible for the press to be forging while arm1 is picking up another metal blank.

The parts should be working concurrently. Here we describe how the basic process will be performed.

1. A blank is put on the feed belt from the stock;
2. the feed belt moves the blank to the table;
3. the table rotates and lifts the blank to a position such that arm1 of the robot can pick it up;
4. arm1 unloads the blank on the press;
5. after being forged by the press [Lin], the blank will be picked up by arm2;
6. arm2 puts the blank on the deposit belt;

[1] http://www.fzi.de/divisions/prost/projects/production_cell/

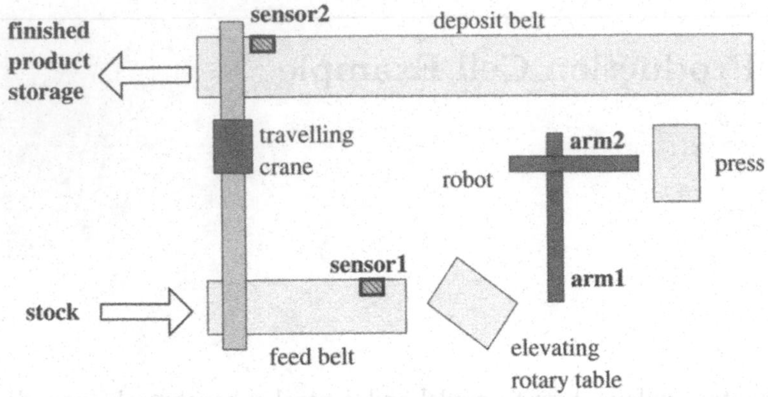

Fig. 5.1. Production Cell

Table 5.1. Modelling the Production Cell by flat DES

	FeedBelt	Table	Robot	Press	DepositBelt	Crane	Total
#automata	1	2	5	1	1	3	13
#states	5	36	2916	9	2	72	6.8×10^8

7. the deposit belt moves the blank to the end point, where a test unit is installed to measure if the forging is successful;
8. the good blanks will be put into finished product storage, while the crane will pick up the unsuccessfully processed blanks and return them to the feed belt to start another cycle.

The main challenge of this example is to prevent collisions among certain parts and at the same time guarantee nonblocking.

The following table demonstrates an attempt to model the production cell example by flat DES models. After adding another 7 automaton specifications, the size of its state space easily exceeds a billion. For such a large system, control design without considering the nonblocking issue is not so difficult because each specification only involves 2-3 interconnected parts. However, with nonblocking an essential requirement and such a large state space to search, nonblocking control design becomes substantially more difficult.

The disadvantage of flat DES modelling is that it does not have a way to express the hierarchical decomposition of the production cell, which may help us better understand the system and speed up the modelling and synthesis process. Fortunately, we find that hierarchical decomposition in the production cell example is natural.

Each subsystem (component) can have different abstract levels of functionality at different levels of modelling. From the top level, i.e, if we just consider the interfaces among the parts, it is easy to see that each component is abstractly a buffer, and the whole system is similar to the Transfer-Line example. For example, the table is a buffer with size 1 if we ignore all the details of its movement except the input/output information, i.e., a blank moving to the table and a blank leaving the table. But in this example, just knowing the most abstract functionality of the table *cannot* answer all questions about the table. For example, in which direction should the table turn after a blank is placed on it? This is a trivial question of course and only needs local information about the table to answer it. But without including this detail in the model, we cannot even pose the question. Also if we look into more details, we can ask some nontrivial questions such as how to prevent collision between arm1 and the press. To answer *all* such questions, no matter how trivial they may seem, one needs to add further detailed information to the model. This, unfortunately, will increase the burden of synthesis with the increase of state space size. In a flat modelling scheme, adding the new information will increase the synthesis burden dramatically as the *search state space size* [2] will definitely be larger. But our symbolic synthesis approach is not overly sensitive to state space size. With STS as a powerful tool to *encapsulate* the details in lower levels, we can take advantage of the structure of the state space to speed up the search.

Our objective is to model *all* necessary details of the benchmark example and answer *all* questions asked in the original document [Lin, Bra], using an "intelligent" approach based on our STS framework. Basically, besides nonblocking, there are two types of requirements on this benchmark example. One is on buffer size and the other on safety. Most buffer size requirements are trivial and will be answered at the modelling stage [3]. Then we will write specifications for more difficult requirements about safety, and design nonblocking controllers for them.

First we describe the STS model of the Production Cell and summarize some modelling heuristics in section 5.2, give its complete specification in section 5.3, then discuss the synthesis in section 5.4, and finally summarize the chapter in section 5.5.

5.2 STS Model of the Production Cell

In this section, we model each component respectively and provide a global view of the plant. For each component, we first try to give a view from the top

[2] The state space which we need to search to get the controller.

[3] Review the control problem of the Small Factory in chapter 4. The requirement about buffer underflow is trivial (disabling a controllable event). For the Production Cell, the only special buffer size requirement is the one for the elevating rotary table, which we do need to perform supervisory control design. We will discuss it in detail in the next section.

level, i.e., the interface of the component; and then give the inner transition structure as well.

5.2.1 Feed Belt

The feed belt transports workpieces to the elevating rotary table. A sensor is located at the end of the belt to signal when a blank enters or leaves. A blank can be put on the belt from the stock or the crane; and the belt will move the blank to the table. Figure 5.2 demonstrates the input/output communications between feed belt and crane, table and stock.

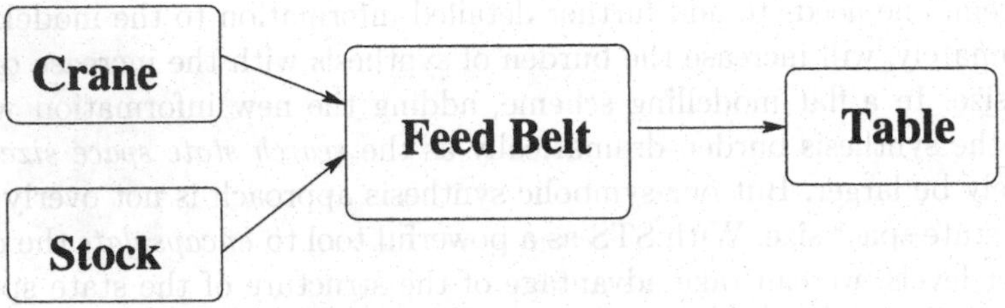

Fig. 5.2. Interface of the Feed Belt

Notice that at most 2 blanks are allowed to be on the feed belt at the same time, because blanks may collide if they are too close on the belt and we have only 1 sensor to tell how far apart they are. So the belt is like *a buffer with size 2 at the top level.* But the inner structure is more complex. We need to know if a blank is at the end or not and the belt is moving or at rest. Three parameters of the belt used to build the state space are shown in Table 5.2.

Table 5.2. Building the state space of Feed Belt

Parameter	Value	Description
#blanks	0,1,2	The belt can take 0, 1, 2 blanks
Sensor	off, on	The sensor decides if 1 blank is at the end
Movement	F, S	The belt can forward or stop

The state is taken to be the triple (#blanks, sensor, movement), e.g., (1,on,F) is the state where 1 blank is at the end of the feed belt(sensor *on*) and the belt is forwarding (F). The state space size is $3 \times 2 \times 2 = 12$. After erasing some trivial unreachable states, we arrive at the following 2-level transition graph FB, shown in Figure 5.3. We can see that without looking into the details, the feed belt is just a buffer with size 2. Each superstate of FB is labelled by the number of blanks on the belt. Notice that our STS is different from Wang's STS [Wan95]. Here we allow the initial(marker) states to be at the lower level. This makes the modelling more flexible.

Table 5.3 lists all events in FB and their properties.

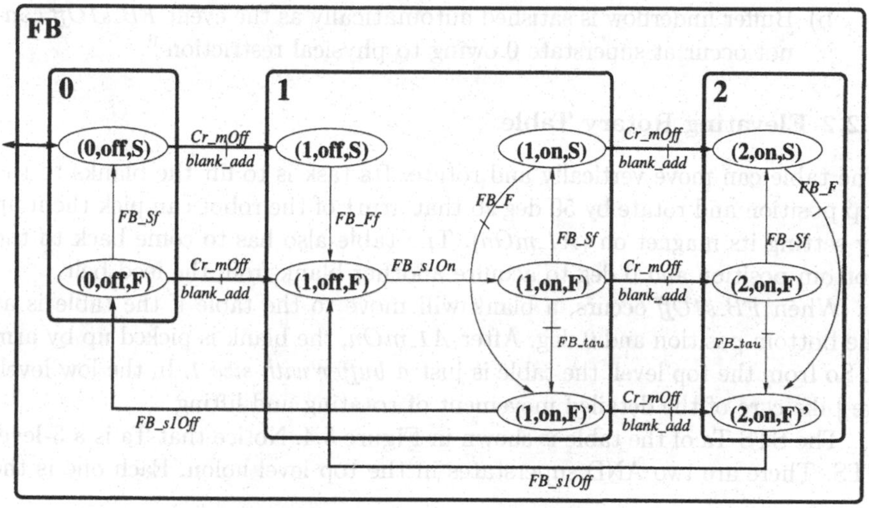

Fig. 5.3. STS of the Feed Belt

Table 5.3. Events in FB

Event	Shared?	Controllable?	Description
Cr_mOff	Y	Y	Crane puts a blank on the belt
FB_s1Off	Y	N	Sensor1 off, a blank leaves the belt
blank_add	N	Y	Add a blank from the stock
FB_F	N	Y	Feed belt forwards
FB_Ff	N	N	Feed belt forwards
FB_Sf	N	N	Feed belt stops
FB_s1On	N	N	Sensor1 on, a blank reaches the end
FB_tau	N	Y	A short period of time has elapsed

Remark:

1. Only the events on the top level could be shared with other subsystems.
2. Notice that in superstate 1, in order to prevent a blank from falling (*FB_s1Off*) when the table is not ready, we introduce a new event *FB_tau*, indicating the elapse of a short period of time, and assume *FB_Sf* is forcible such that *FB_tau* can be preempted. Thus *FB_tau* is controllable [4].

[4] See chapter 9 of [Won04]

3. Some control actions are already modelled in FB. They are valid because all events disabled are controllable [5].
   a) The buffer's overflow requirement is satisfied by disabling *Cr_mOff* and *blank_add* at superstate 2, which is already modelled in FB.
   b) Buffer underflow is satisfied automatically as the event *FB_s1Off* cannot occur at superstate 0 owing to physical restriction [6].

5.2.2 Elevating Rotary Table

The table can move vertically and rotate. Its task is to lift the blanks to the top position and rotate by 50 deg so that arm1 of the robot can pick them up by setting its magnet on (*A1_mOn*). The table also has to come back to the bottom position and 0 deg to acquire another blank from the feed belt.

When *FB_s1Off* occurs, a blank will move to the table if the table is at the bottom position and 0 deg. After *A1_mOn*, the blank is picked up by arm 1. So from the top level, the table is just *a buffer with size 1.* In the low level, we take care of the detailed movement of rotating and lifting.

The STS Ta of the table is shown in Figure 5.4. Notice that Ta is a 3-level STS. There are two AND superstates in the top level holon. Each one is the

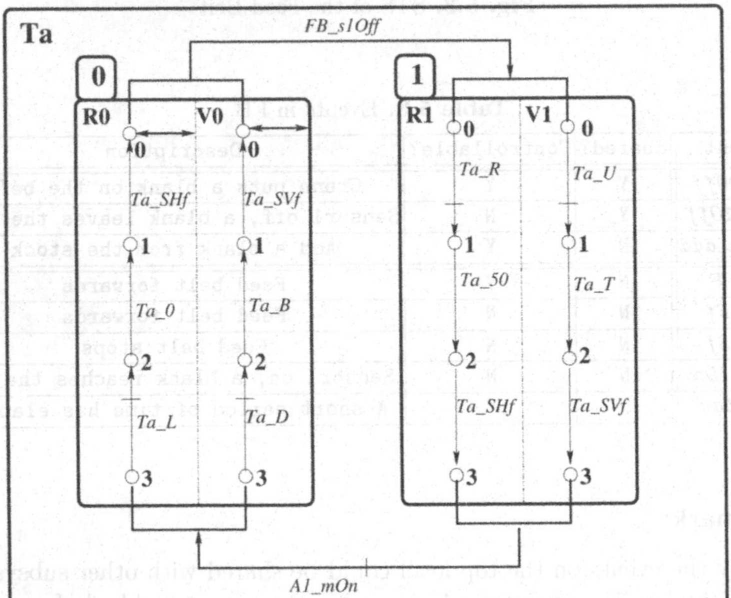

Fig. 5.4. STS of the Table

---

[5] Recall the buffer specification of the Small Factory in chapter 4.

[6] No *blank-leaving* will occur if no blank is on the belt.

synchronous product of two holons modelling rotating and lifting, respectively. Table 5.4 lists all events in Ta and their properties.

Table 5.4. Events in Ta

Event	Shared?	Controllable?	Description
FB_s1Off	Y	N	Feed belt sensor off, a blank comes
A1_mOn	Y	Y	Arm1 picks one blank
Ta_L	N	Y	Table turns left
Ta_R	N	Y	Table turns right
Ta_SHf	N	N	Table stops horizontal movement, forcible
Ta_0	N	N	Table reaches 0 deg
Ta_50	N	N	Table reaches 50 deg
Ta_U	N	Y	Table moves up
Ta_D	N	Y	Table moves down
Ta_SVf	N	N	Table stops vertical movement, forcible
Ta_T	N	N	Table reaches the top
Ta_B	N	N	Table reaches the bottom

The STS Ta gives us a clean description of the table's behavior. The reader should find it easy to understand the behavior. One may try to model the table's behavior by the flat model of synchronous product. However, the result would not be as comprehensible as our STS model.

Remark:

1. The buffer underflow requirement is achieved by disabling controllable event *A1_mOn*. This is already modelled in STS Ta.
2. The buffer overflow requirement will be answered in the synthesis stage. The event *FB_s1Off* is uncontrollable. So we cannot model this requirement into the STS Ta. But we can write a logic specification about it later.

5.2.3 Robot

The robot consists of 2 orthogonal arms. Each one can extend or retract. Each can load or unload a blank by turning its magnet on or off. Both rotate jointly with the robot, which in this case study makes the robot the most difficult part to model.

The robot's task is to transport workpieces from the table to the press by arm1, and then from the press to the deposit belt by arm2. Both can be done simultaneously to maximize the capacity of the press. But care is needed because both arms are connected to the robot and rotate jointly. At the top level the robot is *similar to a buffer with size 2*, while at the low level it is much more complex.

Figure 5.5 demonstrates the interface of the robot.

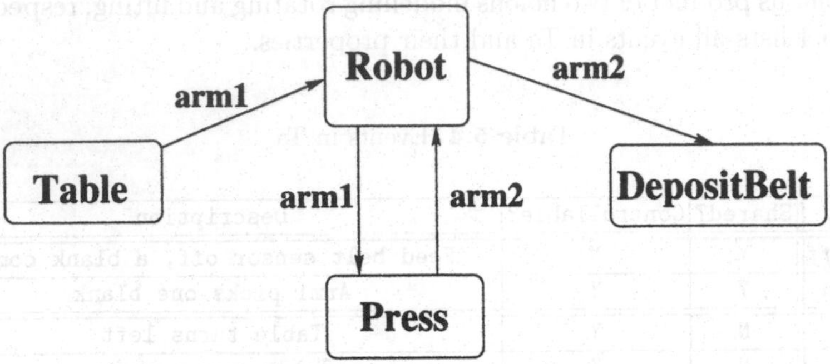

Fig. 5.5. Interface of the Robot

On its page 13, [Lin] gives specific instructions on operating the robot, so as to maximize the capacity of the press. Assume the press has been loaded and arm1 is at the correct position to pick up another blank from table. Then the required order of events will be:

1. arm1 picks up a blank from the table and then the robot starts rotating;
2. arm2 picks up the blank finished by the press when the robot rotates to an assigned angle;
3. arm2 releases the blank on the deposit belt and arm1 releases its blank on the press;
4. robot comes back to the original position to start another cycle.

The robot is required to work in this cyclic fashion. The STS Ro in Figure 5.6 models such behavior.

Table 5.5 lists all events in Ro.

The following is the explanation of the transition graph in Figure 5.6.

1. At state 0 of Ro, the initial state, assume arm1 is pointing to the elevating rotary table and at the right extension to pick up a blank. Also assume arm2 is at safe extension to avoid possible collision between arm2 and press.
2. After arm1 picks up a blank, i.e., *A1_mOn* occurs, superstate 1 is reached. Inside superstate 1, robot rotates to -90 deg(*Ro_-90*) and arm1 extends to the position(*A1_65*) at the same time before arm1 unloads the blank(*A1_mOff*) on the press. Now the press is loaded and the working cycle just described can start at superstate 2.
3. Inside AND superstate 2, on the one hand, arm2 retracts to a safe extension(*A2_0*) to avoid collision with the press, and then the robot turns back to 50 deg(*Ro_50*); at the same time arm1 retracts to the position (*A1_52*) in order to pick up another blank. Now if a blank is ready for pickup on the table, arm1 will be loaded (*A1_mOn*).

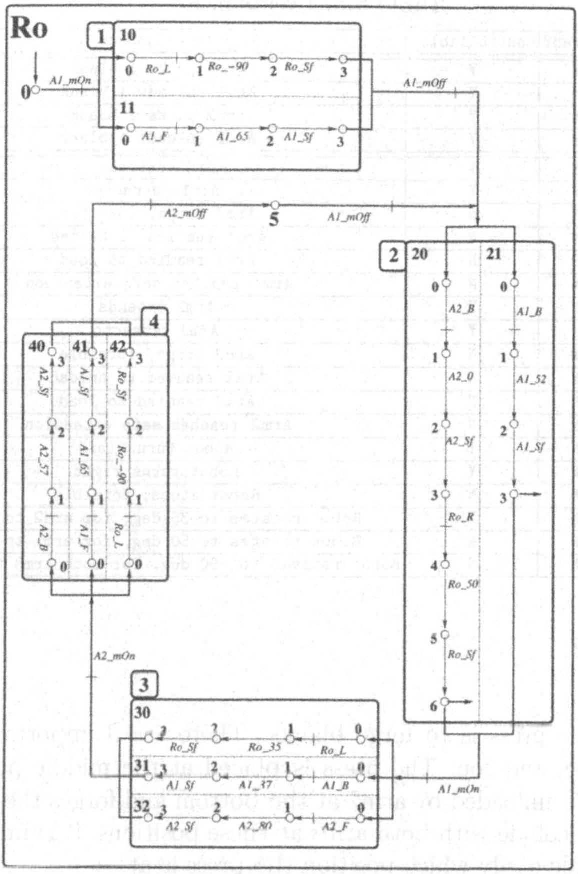

Fig. 5.6. STS of the Robot

4. Inside AND superstate 3, there are three AND components running concurrently. They execute the prerequisites to loading arm2. One holon embodies the fact that the robot needs to rotate to 35 deg(*Ro_35*). Another embodies the fact that arm2 needs to extend to the position (*A2_80*). The last embodies the fact that arm1 needs to retract to *A1_37* for safety as it might otherwise collide with the press. Now if a forged blank is ready for pickup in the press, arm2 will be loaded (*A2_mOn*).
5. Now we need to unload both arms at the correct position. To unload them, robot must be at -90 deg(*Ro_-90*), arm1 at *A1_65* and arm2 at *A2_57*. These requirements are modelled in superstate 4.
6. After that, the system can unload arm2 (*A2_mOff*) and arm1 (*A1_mOff*). The system will then return to superstate 2 for another working cycle.

Table 5.5. Events in Ro

Event	Shared?	Controllable?	Description
A1_mOn	Y	Y	Arm1 picks 1 blank
A1_mOff	Y	Y	Arm1 unloads 1 blank
A2_mOn	Y	Y	Arm2 picks 1 blank
A2_mOff	Y	Y	Arm2 unloads 1 blank
A1_F	N	Y	Arm1 extends
A1_B	N	Y	Arm1 retracts
A1_Sf	N	N	Arm1 stops, forcible
A1_65	N	N	Arm1 readied to unload
A1_52	N	N	Arm1 readied to load
A1_37	N	N	Arm1 reaches safe extension
A2_F	N	Y	Arm2 extends
A2_B	N	Y	Arm2 retracts
A2_Sf	N	N	Arm2 stops, forcible
A2_57	N	N	Arm2 readied to unload
A2_80	N	N	Arm2 readied to load
A2_0	N	N	Arm2 reaches safe extension
Ro_L	N	Y	Robot turns left
Ro_R	N	Y	Robot turns right
Ro_Sf	N	N	Robot stops, forcible
Ro_35	N	N	Robot rotates to 35 deg, for arm2 to load
Ro_50	N	N	Robot rotates to 50 deg, for arm1 to load
Ro_-90	N	N	Robot rotates to -90 deg, for both arms to unload

5.2.4 Press

The task of the press is to forge blanks. There are 3 important positions: bottom, middle, and top. The press is placed at the middle position to be loaded by arm1, unloaded by arm2 at the bottom and forges the blank at the top. As it may collide with both arms at these positions, it is important that the model decide easily which position the press is at.

The press is just *a buffer with size 1* at the top level. Its low level transition is simple too. We model it as a 2-level STS. Figure 5.7 displays the transition graph. Table 5.6 lists all events in Pr.

Table 5.6. Events in Pr

Event	Shared?	Controllable?	Description
A1_mOff	Y	Y	Arm1 puts 1 blank on press
A2_mOn	Y	Y	Arm2 picks 1 blank
Pr_U	N	Y	Press moves up
Pr_D	N	Y	Press moves down
Pr_Sf	N	N	Press stops moving, forcible
Pr_T	N	N	Press reaches top position
Pr_M	N	N	Press reaches middle position
Pr_B	N	N	Press reaches bottom position

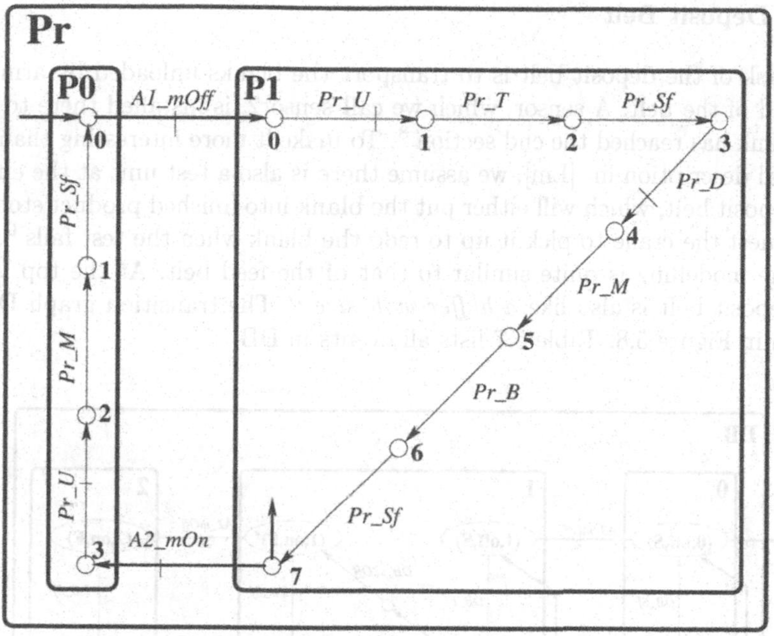

Fig. 5.7. STS of the Press

Here we explain its dynamics. Initially, the press is at the middle position (*Pr_M*) and unloaded (at state 0 of the holon P0). After arm1 unloads a blank on the press (*A1_mOff*), the STS model reaches state 0 inside the superstate P1. Then the press moves to the top position (*Pr_T*) and the blank is forged. After that, the press moves to the bottom position (*Pr_B*) in order for arm2 to pick up the finished blank (*A2_mOn*). Now the STS model is at state 3 of the holon P0. The press will move back to the middle position to wait for another blank.

Notice that state 7 of the holon P1 is the only marker state. The main reason that we do not set the initial state as the only marker state is based on the job description on page 13 of [Lin]. During the working cycle of the robot, at least one blank should be on press or robot. So it is impossible to have both robot and press unloaded [7]. The compromise is necessary in order to get a nonempty solution.

Remark:

1. The buffer overflow requirement is satisfied by disabling *A1_mOff*. The buffer underflow requirement is satisfied by disabling *A2_mOn*. Both are already modelled in Pr.

[7] In that case, **Sync(Pr, Ro)** is empty.

5.2.5 Deposit Belt

The task of the deposit belt is to transport the blanks unloaded by arm2 to the end of the belt. A sensor, which we call sensor2, is installed there to test if a blank has reached the end section [8]. To make it more interesting than the original description in [Lin], we assume there is also a test unit at the end of the deposit belt, which will either put the blank into finished product storage, or request the crane to pick it up to redo the blank when the test fails [9].

The modelling is quite similar to that of the feed belt. At the top level, the deposit belt is also like *a buffer with size 2*. The transition graph DB is shown in Figure 5.8. Table 5.7 lists all events in DB.

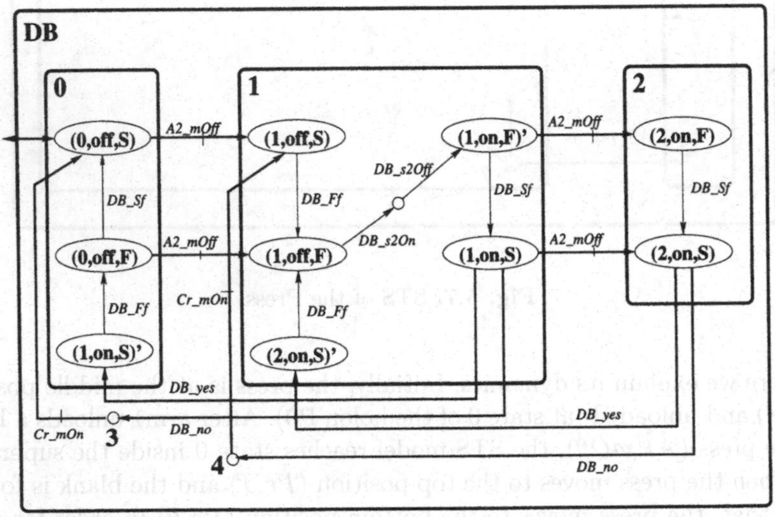

Fig. 5.8. STS of the Deposit Belt

Remark:

1. The buffer overflow requirement is satisfied by disabling *A2_mOff* at superstate 2. The buffer underflow requirement is satisfied by disabling *Cr_mOn* at superstate 0. Both are already modelled in the STS DB.

5.2.6 Crane

The task of the crane is to pick up blanks from the deposit belt, move them to the feed belt, and unload them there. The crane also has an electromagnet

[8] Sensor2 works a little differently from the sensor1 at the feed belt. An event *sensor2 off* following *sensor2 on* indicates a blank is at the end, whereas in sensor1, event *sensor1 on* alone indicates the arrival of a blank.

[9] This is just like the test unit in the Transfer-Line example.

Table 5.7. Events in DB

Event	Shared?	Controllable?	Description
A2_mOff	Y	Y	Arm2 puts 1 blank on deposit belt
Cr_mOn	Y	Y	Crane picks up 1 blank
DB_yes	N	N	Test unit reports yes
DB_no	N	N	Test unit reports no
DB_Ff	N	N	Deposit belt forwards, forcible
DB_Sf	N	N	Deposit belt stops, forcible
DB_s2On	N	N	Sensor2 on, a blank moves to the end
DB_s2Off	N	N	Sensor2 off, a blank reaches the end

and horizontal, vertical mobility. The horizontal motion covers the distance between the belts. The vertical motion is necessary because the belts are placed at different levels.

So from the top level, the crane is just *a buffer with size 1.* In the low level, we take care of the detailed movement.

The STS Cr is shown in Figure 5.9. Table 5.8 lists all events in Cr.

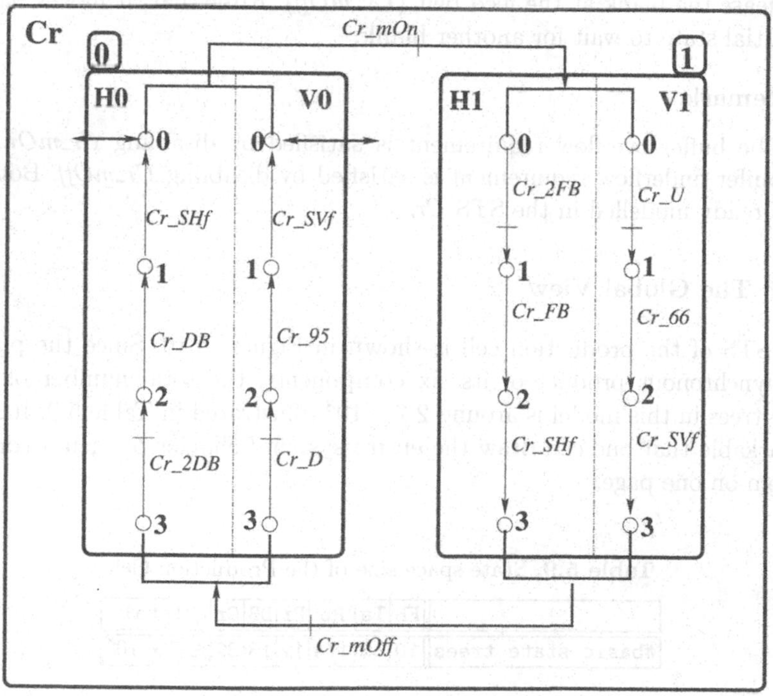

Fig. 5.9. STS of the Crane

Table 5.8. Events in Cr

Event	Shared?	Controllable?	Description
Cr_mOn	Y	Y	Crane picks up 1 blank
Cr_mOff	Y	Y	Crane unloads 1 blank
Cr_U	N	Y	Crane's magnet moves up
Cr_D	N	Y	Magnet moves down
Cr_SVf	N	N	Crane stops vertical movement, forcible
Cr_66	N	N	Magnet's height readied for unloading
Cr_95	N	N	Magnet's height readied for loading
Cr_2FB	N	Y	Crane moves to feed belt
Cr_2DB	N	Y	Magnet moves to deposit belt
Cr_SHf	N	N	Crane stops horizontal movement, forcible
Cr_FB	N	N	Crane reaches feed belt
Cr_DB	N	N	Crane reaches deposit belt

Here we explain its dynamics. Initially, the crane is over the deposit belt (at state 0 of the holon H0) and at the proper height (at state 0 of the holon V0). When it picks up an unsuccessfully processed blank (*Cr_mOn*), the crane moves towards the feed belt and at the same time adjusts its height in order to release the blank at the feed belt (*Cr_mOff*). After that, it moves back to its initial state to wait for another blank.

Remark:

1. The buffer overflow requirement is satisfied by disabling *Cr_mOn*. The buffer underflow requirement is satisfied by disabling *Cr_mOff*. Both are already modelled in the STS *Cr*.

5.2.7 The Global View

The STS of the production cell is shown in Figure 5.10. Since the plant is the synchronous product of its six components, the total number of basic state trees in this model is around 2.7×10^8, illustrated in Table 5.9. It seems remarkable that one can draw the entire system behavior of such a complex system on one page.

Table 5.9. State space size of the Production Cell

	FB	Ta	Ro	Pr	DB	Cr	Total
#basic state trees	10	32	174	12	13	32	2.7×10^8

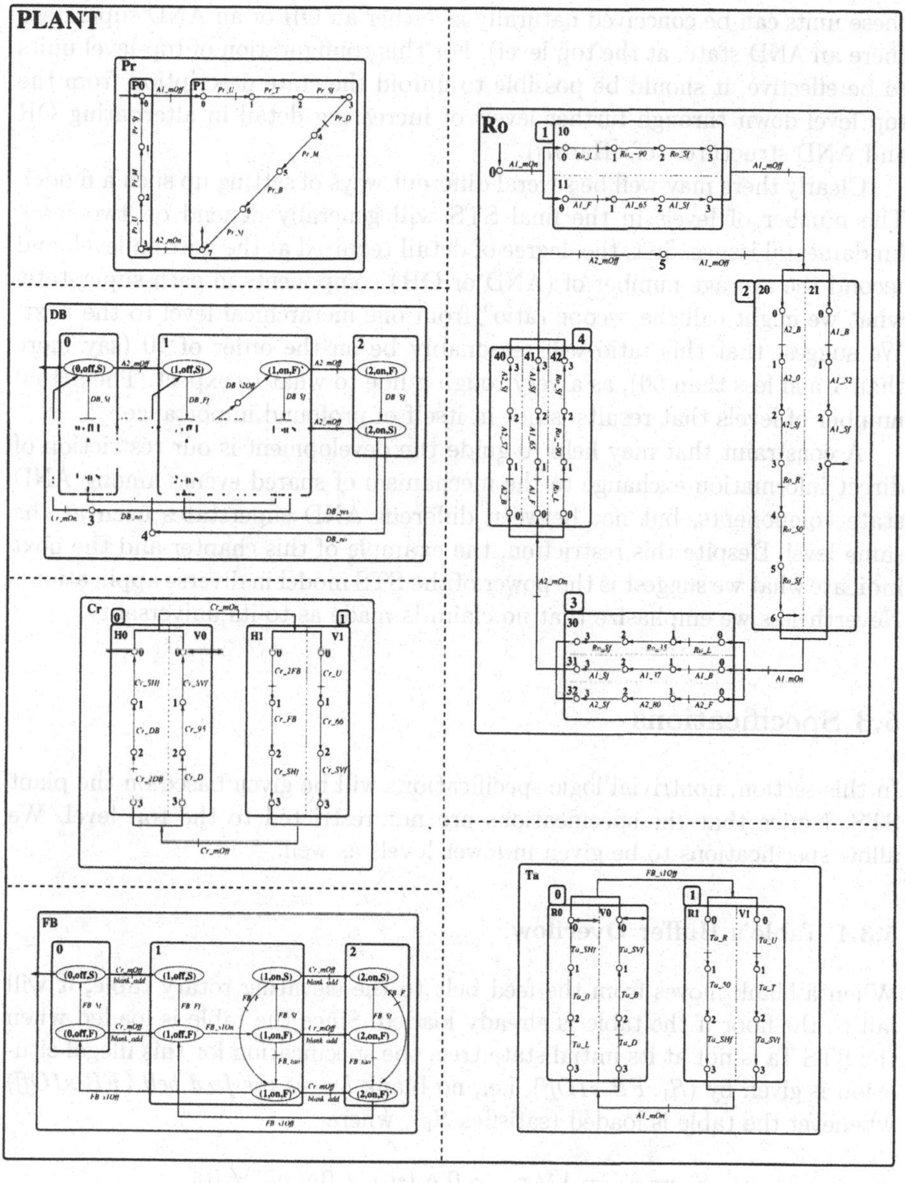

Fig. 5.10. STS of the Production Cell

5.2.8 Modelling Heuristics

The reader may be wondering how STS modelling is done in general. As in the example described in this chapter, we expect that a complex system will be built up from identifiable physical components interacting in specific ways. For STS the objective is first to specify a maximal aggregation of components into top-level units (here the six main units in Fig. 5.10). Our assumption is that these units can be conceived naturally as either an OR or an AND superstate (here an AND state, at the top level). For this configuration of top-level units to be effective, it should be possible to unfold the state description from the top level down through further levels of increasing detail in alternating OR and AND structures (cf. [Har87]).

Clearly there may well be several different ways of setting up such a model. The number of levels in the final STS will generally depend on two more fundamental issues: first, the degree of detail required at the bottom level, and second the average number of (AND or OR) components in each superstate: what we might call the "scope ratio" from one hierarchical level to the next. We suggest that this ratio will reasonably be on the order of 10 (say more than 1 and less than 50), as a very rough guide to what to expect. The actual number of levels that results is not in itself of profound importance.

A constraint that may help to guide the development is our restriction of direct information exchange to the mechanism of shared events among AND state components, but not between different AND superstates even at the same level. Despite this restriction, the example of this chapter and the next indicate what we suggest is the power of the STS model in diverse applications. Nevertheless we emphasize that no claim is made as to its universality.

5.3 Specifications

In this section, nontrivial logic specifications will be given based on the plant STS. Notice that the specifications are not restricted to the top level. We allow specifications to be given in lower levels as well.

5.3.1 Table's Buffer Overflow

When a blank moves from the feed belt to the elevating rotary table, it will fall to the floor if the table is already loaded. Since the table is loaded when the STS Ta is not at its initial state tree, the specification for this illegal situation is given by $(S_1, \mathit{FB\_s1Off})$, i.e., no blank *leaves the feed belt* ($\mathit{FB\_s1Off}$) whenever the table is loaded (satisfies S_1), where

$$S_1 := v_{\text{Ta}} = 1 \vee v_{\text{Ta}} = 0 \wedge (v_{\text{Ro}} \neq 0 \vee v_{\text{Vo}} \neq 0).$$

5.3.2 Collision between Arm1 and Press

Arm1 and press may collide if either of the following two sets of conditions holds.

1. Inside the superstate 2 of STS Ro and the superstate P1 of STS Pr, right after arm1 puts a blank on the press, collision will occur if the press starts moving up to the top position at state 1 of holon P1 and arm1 is still

inside the press. As *arm1 may still be inside the press* before the robot rotates to 50 deg (*Ro_50*), the dangerous states are $\{0, 1, 2, 3, 4\}$ of the holon 20 in the STS Ro. So the specification is given by

$$S_2 := (v_{\mathrm{Pr}} = \mathrm{P1} \wedge v_{\mathrm{P1}} = 1) \wedge (v_{\mathrm{Ro}} = 2 \wedge v_{20} \in \{0, 1, 2, 3, 4\}).$$

2. Inside superstate 4 of STS Ro and superstate P1 of STS Pr, when arm1 is approaching the press, collision will occur if the press is at the top position, arm1 extends over 0.37 (event *A1_37*), and the robot is at -90 deg. The press is at the top position when the holon P1 of STS Pr is at $\{2, 3, 4\}$. Arm1 extends over 0.37 when the holon 41 of STS Ro is at $\{1, 2, 3\}$. The robot is at -90 deg when the holon 42 of STS Ro is at $\{2, 3\}$. So the specification is given by

$$S_3 := (v_{\mathrm{Pr}} = \mathrm{P1} \wedge v_{\mathrm{P1}} \in \{2, 3, 4\}) \wedge (v_{\mathrm{Ro}} = 4 \wedge v_{41} \in \{1, 2, 3\} \wedge v_{42} \in \{2, 3\}).$$

5.3.3 Collision between Arm2 and Press

Arm1 and press may collide if either of the following two sets of conditions holds.

1. Inside superstate 4 of holon Ro and superstate P0 of holon Pr, right after arm2 picks up a blank from the press, collision will occur if the press starts moving up to the middle position at state 2 of holon P0 and arm2 is still inside the press. As arm2 is still inside the press before the robot rotates to -90 deg (*Ro_-90*), the dangerous states are $\{0, 1\}$ of the holon 42 in STS Ro. So the specification is given by

$$S_4 := (v_{\mathrm{Pr}} = \mathrm{P0} \wedge v_{\mathrm{P0}} = 2) \wedge v_{\mathrm{Ro}} = 4 \wedge v_{42} \in \{0, 1\}.$$

2. Inside superstate 3 of holon Ro, when arm2 is approaching the press, collision will occur if the press is at the middle position, arm2 extends over 0, and robot is at 35 deg. The press is at the middle position when holon P0 is at $\{0, 1\}$ or holon P1 is at $\{0, 1, 5\}$. Arm2 extends over 0 when the holon 32 of STS Ro is at $\{1, 2, 3\}$. The robot is at 35 deg when holon 30 of STS Ro is at $\{2, 3\}$. So the specification is given by

$$S_5 := (v_{\mathrm{Pr}} = \mathrm{P0} \wedge v_{\mathrm{P0}} \in \{0, 1\} \vee v_{\mathrm{Pr}} = \mathrm{P1} \wedge v_{\mathrm{P1}} \in \{0, 1, 5\}) \wedge$$
$$\wedge (v_{\mathrm{Ro}} = 3 \wedge v_{32} \in \{1, 2, 3\} \wedge v_{30} \in \{2, 3\}).$$

5.3.4 Collision between Arm1 and Table

The table and arm1 may collide if both hold a blank, arm1 is still pointing at the table and the table is at the top position. Then the two blanks can touch each other, which is detected as a collision. This scenario will unfold after arm1 is loaded but the robot has not turned in order to avoid collision between arm2 and the press. Now if the table takes another blank and lifts it to the top position, the collision will occur.

Since this will happen only when STS Ro is at superstate 3 and STS Ta is at superstate 2, we can write down its conditions as follows,

1. holon 30 of STS Ro is at $\{0,1\}$ when arm1 is still pointing at the table;
2. holon 31 of STS Ro is at $\{0,1\}$ when arm1 has not retracted to its safe extension;
3. holon V1 of STS Ta is at $\{2,3\}$ when the table is at the top position.

So the illegal predicate is given by

$$S_6 := (v_{\mathrm{Ro}} = 3 \wedge v_{30} \in \{0,1\} \wedge v_{31} \in \{0,1\}) \wedge (v_{\mathrm{Ta}} = 1 \wedge v_{\mathrm{V1}} \in \{2,3\}).$$

5.4 Nonblocking Supervisory Control Design

The Production Cell example is moderately complex, with state space of order 10^8 and 6 logic formulas as its specification. Our BDD-based program can automatically compute the controller in less than 5 seconds on a personal computer with 1G Hz Athlon CPU and 256MB RAM.

This chapter only focuses on how to describe a complex control problem. We will discuss the synthesis in detail when we investigate a larger example in Chapter 6.

5.5 Summary

The production cell example has some interesting properties, of which the most noteworthy are summarized here.

1. It has a hierarchical decomposition. Information hiding is an important requirement for us to exploit the STS framework.
2. It is difficult. There are many specifications, including trivial and nontrivial ones. So it would probably be infeasible to design a controller using flat DES models.

The most important advantage of our STS framework demonstrated in this chapter is that it can provide users with an integrated view of the control problem. So compared with the modular control methodology, the STS framework ought to be more readily accepted by engineers not specialist in DES.

Another advantage of our STS framework is its efficiency of control design, as we will demonstrate in chapter 6.

6

The AIP Example

6.1 Introduction

The AIP example was first presented by Brandin *et al.* in [BC94]. The whole system was modelled as the synchronous product of 100 automata. Modular control was applied to the synthesis. The AIP system is illustrated in Figure 6.1. There are five conveyor loops. The central loop (CL) communicates with each external loop (L$i, i = 1, 2, 3, 4$) by a transfer unit (TU$i, i = 1, 2, 3, 4$), respectively. There are three assembly stations AS$i, i = 1, 2, 3$ linked to the external loops L$i, i = 1, 2, 3$. Each assembly station AS$i, i = 1, 2$ can perform a unique task (task1 and task2 respectively). But AS3 can perform both tasks.

The specification will be given in detail in section 6.3. Here we just describe the basic functionality of AIP. The I/O station connected to L4 allows the input and removal of pallets in the AIP system. There are two types of pallets to be processed by the assembly stations. Pallets of type 1 must go to AS1 first to undergo task1 and then to AS2 to undergo task2 before leaving the AIP from L4. The pallets of type 2 must instead undergo task2 before task1. In normal operation, pallets are only distributed to L1 and L2 to be processed by AS1 and AS2. However, AS1 (AS2) can be broken down. In that case, the unprocessed pallets, originally shipped to AS1 (AS2), will be redirected to L3 to be processed by AS3 (assumed always operational), which can perform either task. After the down station is repaired, it will go back to the normal routine.

Leduc *et al.* [LLW01c] successfully applied *Hierarchical Interface-based Supervisory Control* theory to the AIP example. That program could automatically synthesize a controller for a simplified version, where the maximum allowed number of pallets on external loop 1 or 2 was limited to be one (the original requirement was ten). However, without using symbolic computation, the program exceeded memory when buffer size was increased from 1 to 2.

Using an approach based on Petri nets, Ghaffari *et al.* [GRX03] dealt with various safety problems for a similar example, designated AIP-RAO. However, their work did not touch on the much more difficult property of nonblocking.

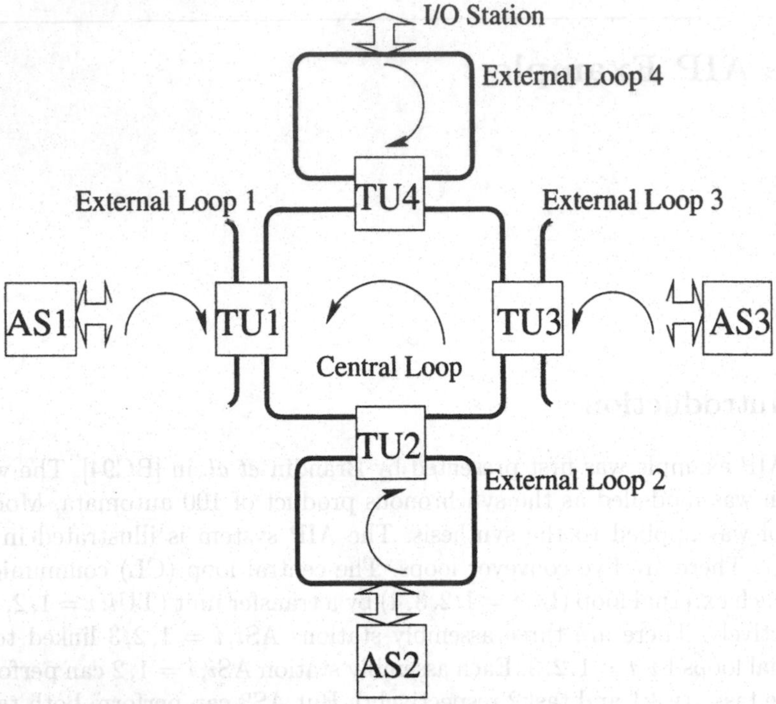

Fig. 6.1. Diagram of AIP

A special case of nonblocking is "reversibility" in the terminology of Petri nets, but reversibility was not addressed either. In this chapter, we will ensure both safety and nonblocking.

We model the AIP example with Brandin's original requirement as a STS with state space of order 10^{24}. Our BDD-based program can automatically compute the controller in less than 20 seconds, using a fixed allocation of memory (required by the BDD package we use) on a personal computer with 1G Hz Athlon CPU and 256MB RAM.

In this chapter, we first describe the STS model of AIP in section 6.2 and its complete specification in section 6.3, then discuss the synthesis in section 6.4, and close with a summary in section 6.5.

6.2 STS Model of AIP

The modelling of AIP is based on the task description given in [BC94, LLW01a]. The interested reader is invited to read Leduc's 76 page report [LLW01a], available on the internet, for a complete description of the system behavior.

The layout of the assembly stations AS$i, i = 1, 2, 3$ is illustrated in Figure 6.2 (from [LLW01a]). Each assembly station can be looked on as a buffer

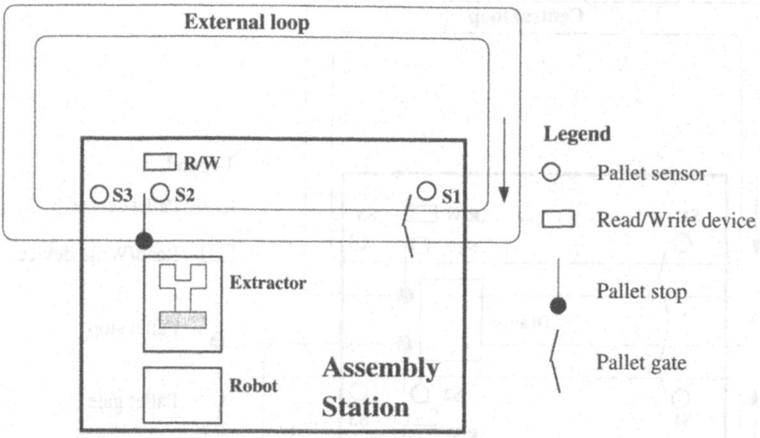

Fig. 6.2. Layout of an assembly station from [LLW01a]

of size 1 at the most abstract level. On looking into the details, one finds more information: there are three important sensors, S1, S2 and S3, to locate the positions of each pallet being processed in the station ; the arrival of a pallet is detected by S1; the pallet gate will be kept closed unless the station is idle at that time; the sensor S2 will be on when a pallet arrives at the pallet stop; if the robot ($ASi, i = 1, 2$) is broken, the pallet will be allowed to pass the stop to be redirected to AS3, otherwise it will be taken by the extractor to feed the robot; after the robot's processing is complete, the read/write device will write information on the label attached to the pallet about the tasks being processed on the pallet; the departure of a pallet is sensed by the sensor S3 and another cycle can start.

The layout of the transfer units $TUi, i = 1, 2, 3, 4$ is illustrated in Figure 6.3 (from [LLW01a]). Each transfer unit can also be looked on as a buffer of size 1 at the top level. The drawer is capable of moving pallets between the central conveyor loop and the external loop. Similar to the layout of assembly stations, on each loop there is exactly the same arrangement of gate, stop and sensors to control access to the unit and locate the pallet inside the unit. Pallets can access the transfer unit from the external loop or the central loop. So we need to control both gates in Figure 6.3 to allow only one pallet in the transfer unit at a time. If a pallet is fed to the unit from the external loop, it must be taken by the drawer to the central loop as it has been processed by the assembly station on that external loop. If a pallet comes from the central loop, it may not be taken to the external loop under certain conditions, e.g., the robot on the external loop is down. The read/write device is placed in the unit to help decide whether the pallet is going to be distributed to that external loop.

Now we proceed to model the AIP. The modelling philosophy is top-down. First designate by AIP the root superstate and then add more details on looking deeper. A rough top level state tree is given in Figure 6.4, according to

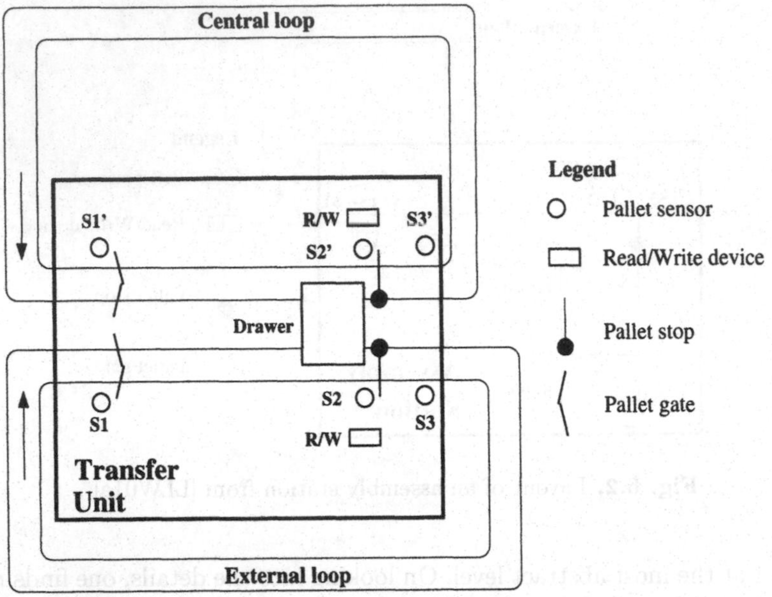

Fig. 6.3. Layout of a transfer unit from [LLW01a]

the diagram of AIP in Figure 6.1. It is an AND decomposition of 9 subsystems. We are already familiar with ASi, $i = 1, 2, 3$ and TUi, $i = 1, 2, 3, 4$. The remaining two, L1 and L2, are so-called *memories* to describe the buffer size of external loop 1 and 2, respectively.

We are going to describe the behavior of each subsystem using STS model. Then we simply build the overall STS model by looking at each subsystem's STS as a child STS of AIP.

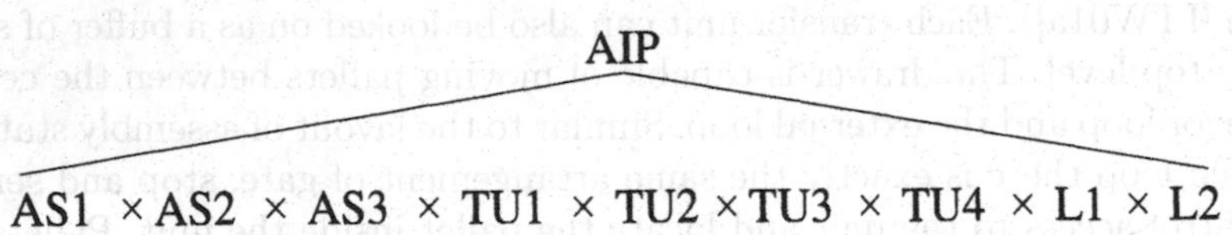

Fig. 6.4. A rough decomposition of the root state AIP

6.2.1 AS1, AS2

Based on the layout of assembly stations in Figure 6.2, we can draw the STS model AS1 in Figure 6.5 and add more low level information for the superstate AS1.Work later in Figure 6.6.

There are three AND components in AS1. The AS1.Robot concerns the robot status. The robot can be broken down and repaired. The AS1.Feed is about feeding pallets at the pallet gate. After S1 is on, we can open the gate if no other pallet is being processed. The event *AS1.gate.open* is controllable, which allows us to avoid the buffer overflow of AS1. The AS1.Process is about processing the pallet. At state 1 of AS1.Process, we can choose two routes. One is to close the stop and process the pallet. Another is to open the stop and let it pass in case the robot is down.

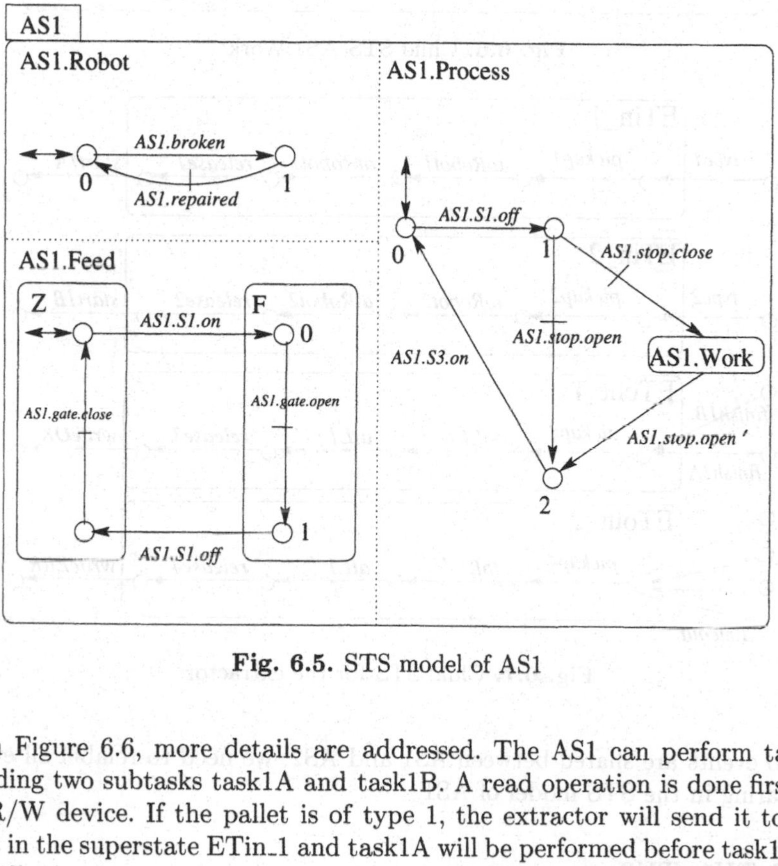

Fig. 6.5. STS model of AS1

In Figure 6.6, more details are addressed. The AS1 can perform task1, including two subtasks task1A and task1B. A read operation is done first by the R/W device. If the pallet is of type 1, the extractor will send it to the robot in the superstate ETin_1 and task1A will be performed before task1B. If the pallet is of type 2, task1B will be performed before task1A. If the tasks are done successfully, the extractor will put the pallet back to the external loop L1 and the R/W device writes *OK* for task1. If any timeout happens during the operation (say, when the robot is down), an *ERROR* will be written on the pallet.

The 4 superstates in AS1.Work are about shipping pallets in and out of the robot. The complexity of their inner transition structures depends on how much detailed information one wishes to add. Here we adopt the simple version of Figure 6.7.

The STS AS1 has 256 basic state trees. The STS model of AS2 will be exactly the same as that of AS1 except for the necessary state/event relabelling.

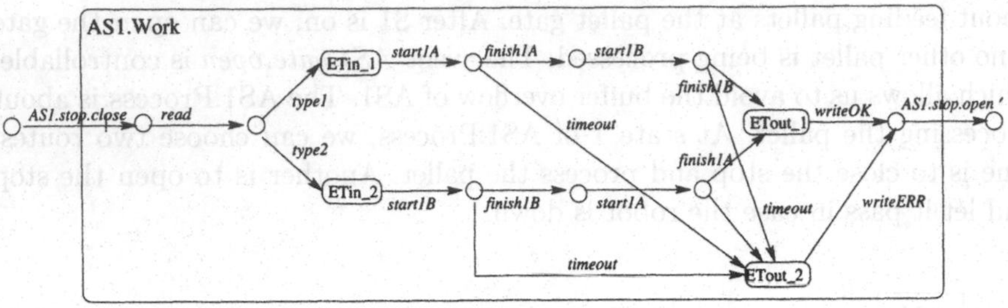

Fig. 6.6. Child STS AS1.Work

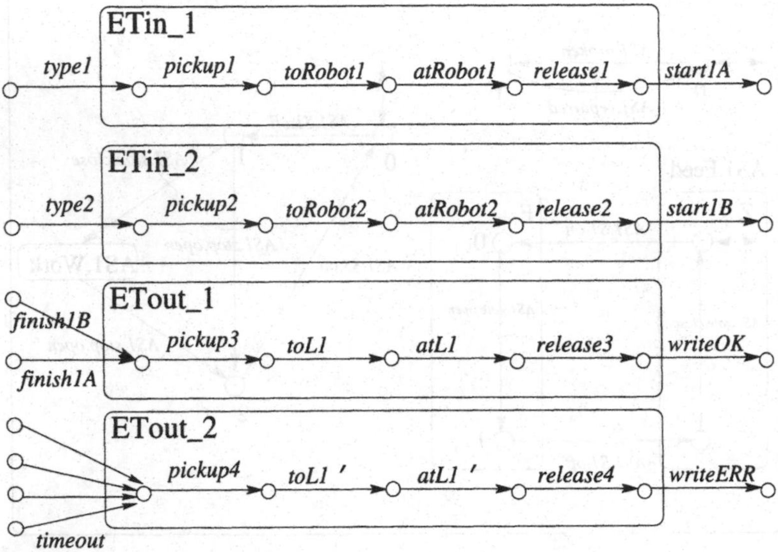

Fig. 6.7. Child STS for the extractor

As no events are shared between AS1 and AS2, we need to relabel all events appearing in the STS model of AS1.

6.2.2 TU1, TU2

Based on the layout of transfer units in Figure 6.3, we depict the STS model TU1 as in Figure 6.8 and add more low level information for the superstates later.

There are four AND components in TU1. The left two in the figure take care of transferring pallets from L1 to CL, whereas the right two are for the transfer of pallets from CL to L1. The L1toCL.Feed and CLtoL1.Feed appear for the same reason as AS1.Feed in AS1. The L1toCL.Transfer actually moves pallets from L1 to CL. The TU1 unconditionally moves a pallet to CL because the pallet is known to come from AS1. However, we need to model a testing step in CLtoL1.Transfer. The CLtoL1.Transfer concerns transferring a pallet to L1 when certain conditions are met. At state 1 of CLtoL1.Transfer, we can choose between two routes. One is to close the stop on CL and process the pallet. The other is to open the stop on CL and let it pass in case the robot of AS1 is down. In the superstate Test&Move, the label on the pallet is going to be read by the R/W device and a decision is made from the information on the label. If the pallet is already being successfully processed by AS1 (notice that the pallets undergoing task1 may come back to TU1 again from the central loop), the stop on CL will be open (*CL.stop.open'*) and will let the pallet pass. In that case transfer fails. Otherwise the drawer will take the pallet to L1 and the stop on L1 will be open (*L1.stop.open*). Now the transfer succeeds.

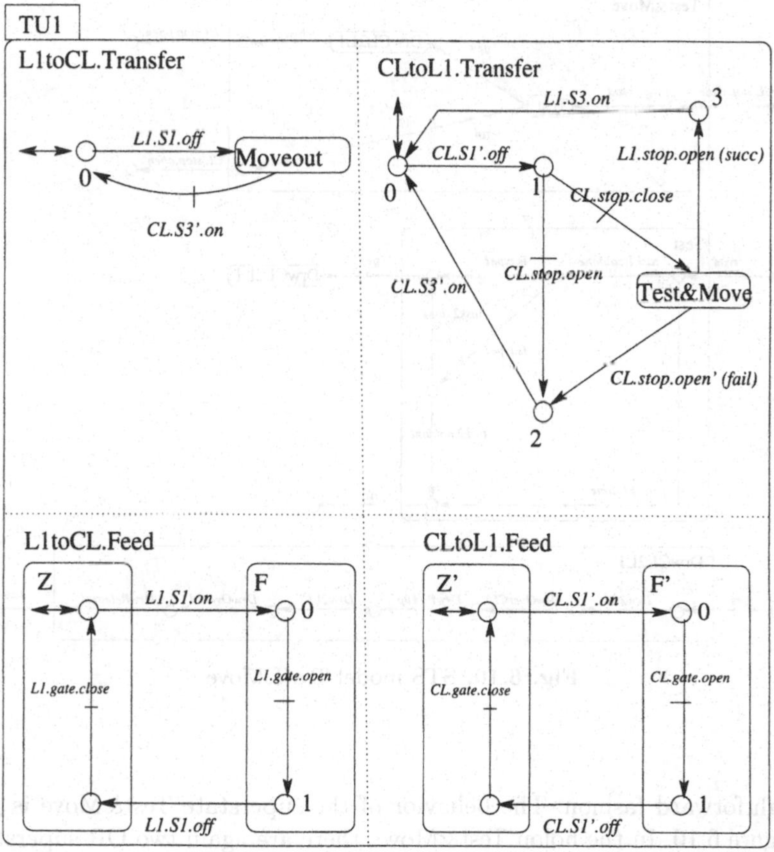

Fig. 6.8. STS model of TU1

There are two superstate children in Figure 6.8, the superstate Moveout and the superstate Test&Move. The behavior of the superstate Moveout is shown in Figure 6.9. As any pallet coming from AS1 at L1 will be unconditionally shipped to CL, the dynamics of Moveout is simple. Inside the superstate Drawer, the detailed information of the drawer's action is described in

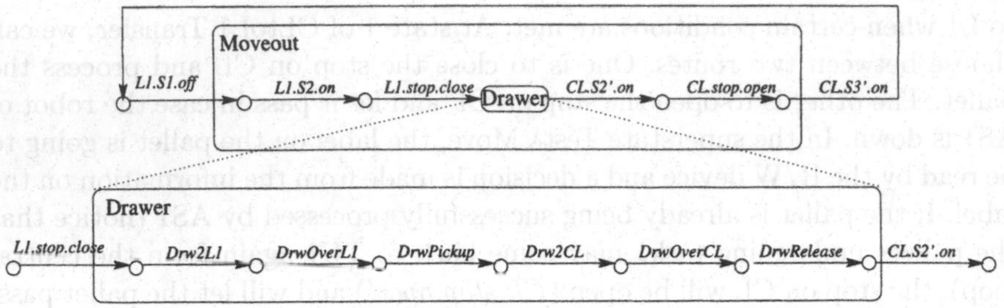

Fig. 6.9. STS model Moveout

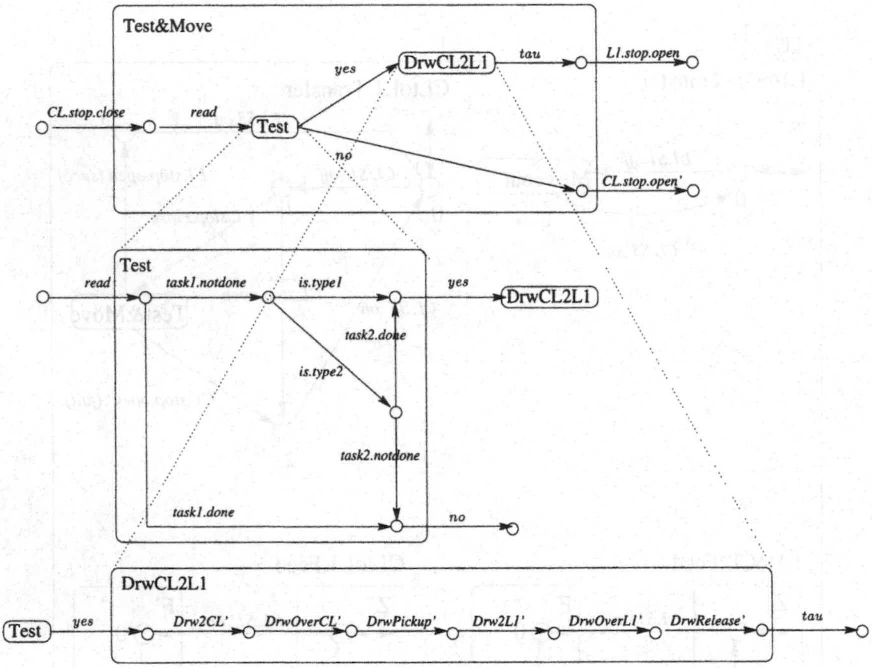

Fig. 6.10. STS model Test&Move

straightforward fashion. The behavior of the superstate Test&Move is given in Figure 6.10. In the holon Test&Move, there are again two OR superstates, Test and DrwCL2L1. The drawer moves the pallet from CL to L1 in the child STS DrwCL2L1. The condition described in the holon Test requires transferring pallets from CL to L1 if and only if

1. the pallet is of type 1 and task1 isn't performed on the pallet, or
2. the pallet is of type 2 and task2 is done and task1 isn't done.

The event *tau*, indicating the elapse of a short period of time, is introduced to arrange that all boundary states are simple states.

The STS TU1 has 3458 basic state trees. The STS model of TU2 will be exactly the same as that of TU1 except for the necessary state/event relabelling.

6.2.3 AS3

AS3 is the backup station for both AS1 and AS2. So its behavior is slightly more complicated. Basically, it has to do the following,

1. repair the damaged pallets assembled by AS1 or AS2;
2. take up the workload of AS1 when AS1 is down;
3. take up the workload of AS2 when AS2 is down.

All the above functions are described in our STS model of AS3, shown in Figure 6.11. As AS3 is unbreakable, we have only two child STS of AS3, where AS3.Feed is about feeding pallets to AS3 and AS3.Process is for the detailed processing of pallets in the station.

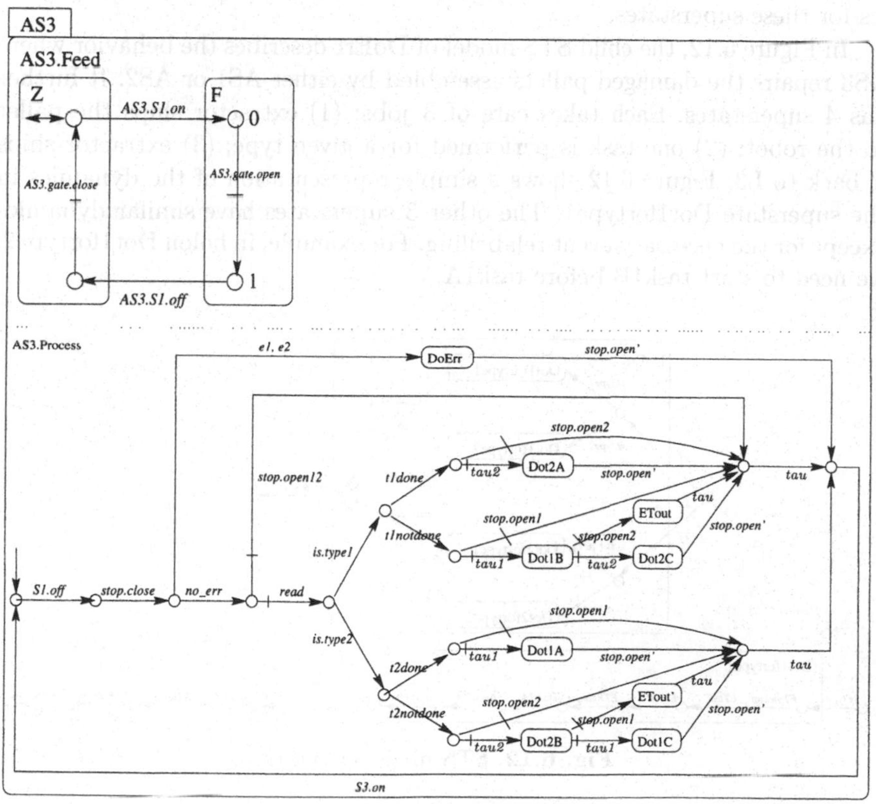

Fig. 6.11. STS model of AS3

Now we explain the behavior in AS3.Process. First a pallet is located at the stop after $S1.off$ [1] occurs. Then the stop must be closed (*stop.close*) and the R/W device reads the label. If the pallet is wrongly assembled by AS1 (event $e1$) or by AS2 (event $e2$), the maintenance is done in the superstate DoErr before the pallet leaves the station ($S3.on$). If no errors are found, the pallet will either pass the stop (*stop.open*12) in case both AS1 and AS2 are up already, or be processed by AS3 otherwise (*read* by R/W device again). The pallet is handled according to its type and status. For example, if it is type 1 and task1 has been performed on it, AS3 will either let it pass the stop directly (by *stop.open2*) if AS2 is already repaired, or otherwise perform task2 in the superstate Dot2A. If it is type 1 and task1 has not yet been performed, AS3 will either let it pass without any processing (by *stop.open1*), or will perform task1 in the superstate Dot1B if AS1 is down. After that, the pallet will be either moved out in ETout immediately, or will undergo task2 in Dot2C if AS2 is also down. Both events *tau1* and *tau2* represent a short period of time that can be preempted by opening the stop, and therefore are controllable.

There are 9 superstates in AS3.Process. We can add more detailed dynamics for these superstates.

In Figure 6.12, the child STS model of DoErr describes the behavior where AS3 repairs the damaged pallets assembled by either AS1 or AS2. It further has 4 superstates. Each takes care of 3 jobs: (1) extractor ships the pallet to the robot; (2) one task is performed for a given type; (3) extractor ships it back to L3. Figure 6.12 shows a simple representation of the dynamics in the superstate Dot1fortype1. The other 3 superstates have similar dynamics except for the necessary event relabelling. For example, in holon Dot1fortype2, we need to start task1B before task1A.

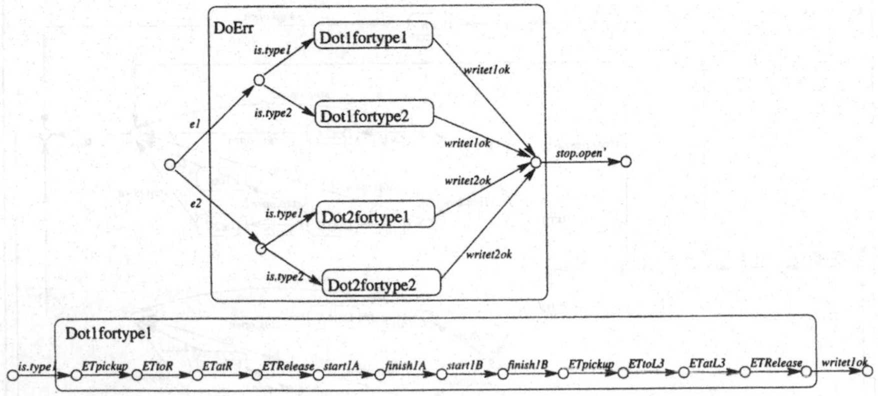

Fig. 6.12. STS model of DoErr

[1] Owing to limited space, we omit the prefix $AS3.$ from all of the events in the graph.

The child STS models of Dot1A, Dot1B, Dot1C and ETout are given in Figure 6.13. Notice that the event set of each child STS is unique. But to save space, we simplify the event labels. For example, the event *start1A* in the child STS Dot1A of Figure 6.13 is the simplified version of the event label *Dot1A.start1A*. Similarly, we can draw the child STS model of Dot2A, Dot2B, Dot2C and ETout′.

AS3 has 2912 basic state trees.

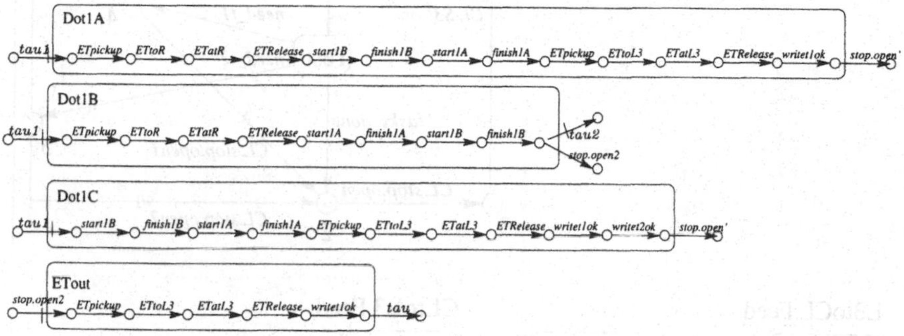

Fig. 6.13. STS model of Dot1A, Dot1B, Dot1C and ETout

6.2.4 TU3

Just like TU1 and TU2, TU3 moves pallets between L3 and CL. Figure 6.14 shows the STS model of TU3.

The behavior of transferring from L3 to CL (see L3toCL.Transfer and L3toCL.Feed in the STS model of TU3) is the same as that in any other transfer units except for the necessary state/event relabelling, because moving out processed pallets from L3 to CL is also unconditional. However, the conditions we require to move a pallet from CL to L3 are different because the robot in AS3 has additional jobs to do. Based on AS3's job description, we need TU3 to move a pallet from CL to L3 if and only if

1. the pallet is wrongly assembled by AS1 or AS2, or
2. task1 needs to be performed while AS1 is down, or
3. task2 needs to be performed while AS2 is down.

Otherwise the pallet will not be transferred. The three conditions are easily modelled into CLtoL3.Transfer. There are two superstates in CLtoL3.Transfer, Test and DrwCL2L3. In Test, the R/W device decides what next task has to be performed on the pallet. For example, if the pallet is of type 1 and only task1 is done, then we need to do task2 (event *need_t2*). In DrwCL2L3, the drawer is going to transfer the pallet. So the holon CLtoL3.Transfer says that

1. at state 1, if the pallet is wrongly assembled by AS1 (event *e*1) or by AS2 (event *e*2), the pallet will be transferred in DrwCL2L3;

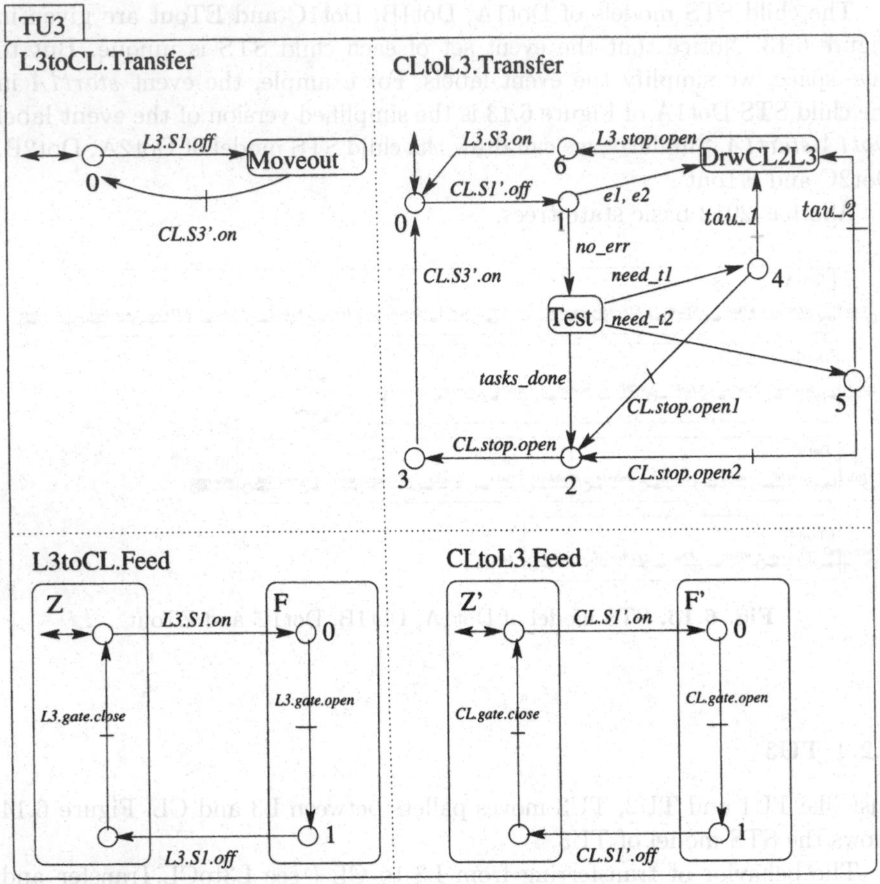

Fig. 6.14. STS model of TU3

2. if task1 needs to be done (*need_t*1), at state 4, the supervisor can choose *tau_*1 to transfer the pallet if AS1 is down, or *CL.stop.open*1 to pass the pallet without transferring if AS1 is working correctly;
3. similarly, if task2 needs to be done (*need_t*2), at state 5, the supervisor can choose *tau_*2 if AS2 is down or *CL.stop.open*2 if AS2 is working correctly.

The STS models of Test and DrwCL2L3 are in turn given in Figure 6.15. TU3 has 6688 basic state trees.

6.2.5 TU4

TU4 moves pallets between L4 and CL. Figure 6.16 shows the STS model of TU4.

The behavior of transferring a pallet from L4 to CL (see L4toCL.Transfer and L4toCL.Feed in the STS model of TU4) is the same as that in any other transfer unit except for the necessary state/event relabelling, because moving

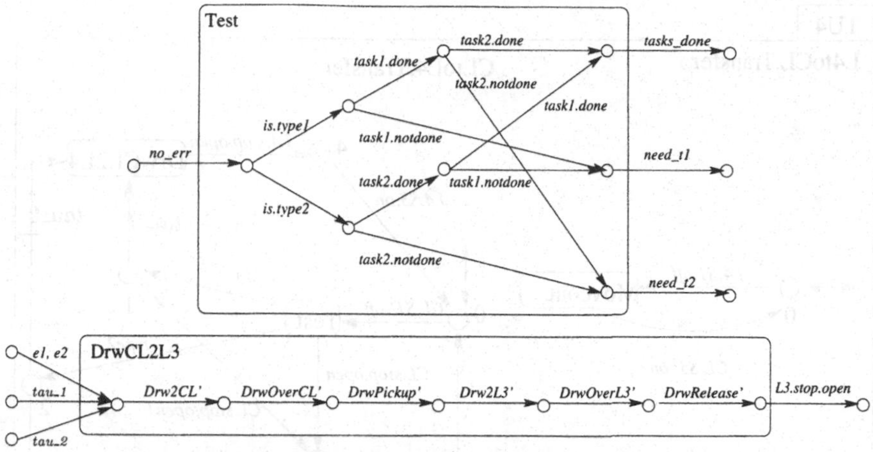

Fig. 6.15. STS model of Test and DrwCL2L3

new pallets from L4 to CL is also unconditional. However, the conditions we require to move a pallet from CL to L4 are different because the pallets are required to leave AIP in a pre-defined order, i.e., one pallet of type 1 and one pallet of type 2, then again one pallet of type 1 and one pallet of type 2, and so on. So we need TU4 to move a pallet from CL to L4 if and only if the pallet type is different from the pallet that is just transferred. Otherwise the pallet will not be transferred. The condition can be easily modelled into CLtoL4.Transfer. There are two superstates in CLtoL4.Transfer, Test and DrwCL2L4. In Test, the R/W device checks if both tasks (1 and 2) are performed on the pallet and then reads its type. There are only three possibilities:

1. both tasks are done and it is of type 1; or
2. both tasks are done and it is of type 2; or
3. at least one task is not done yet.

At state 1 and 2 of CLtoL4.Transfer, we need to decide if it should be transferred by giving a specification later. If at least one task is not done, the stop at CL will open (*CL.stop.open*) and let the pallet pass. So the transfer fails. In DrwCL2L3, the drawer is going to transfer the pallet.

The STS models of Test and DrwCL2L4 are in turn given in Figure 6.17.

TU4 has 4864 basic state trees.

6.2.6 Capacity of L1 and L2

The number of pallets allowed on the out conveyor L1(L2) cannot exceed 10. As shown in Figure 6.18, pallets arrive L1(L2) from either the central loop CL or the assembly station AS1 (AS2). Now we can give the holon L1 that describes the capacity of L1 in Figure 6.19.

The STS model for the capacity of L2 is exactly the same as that of L1 except for the necessary state/event relabelling.

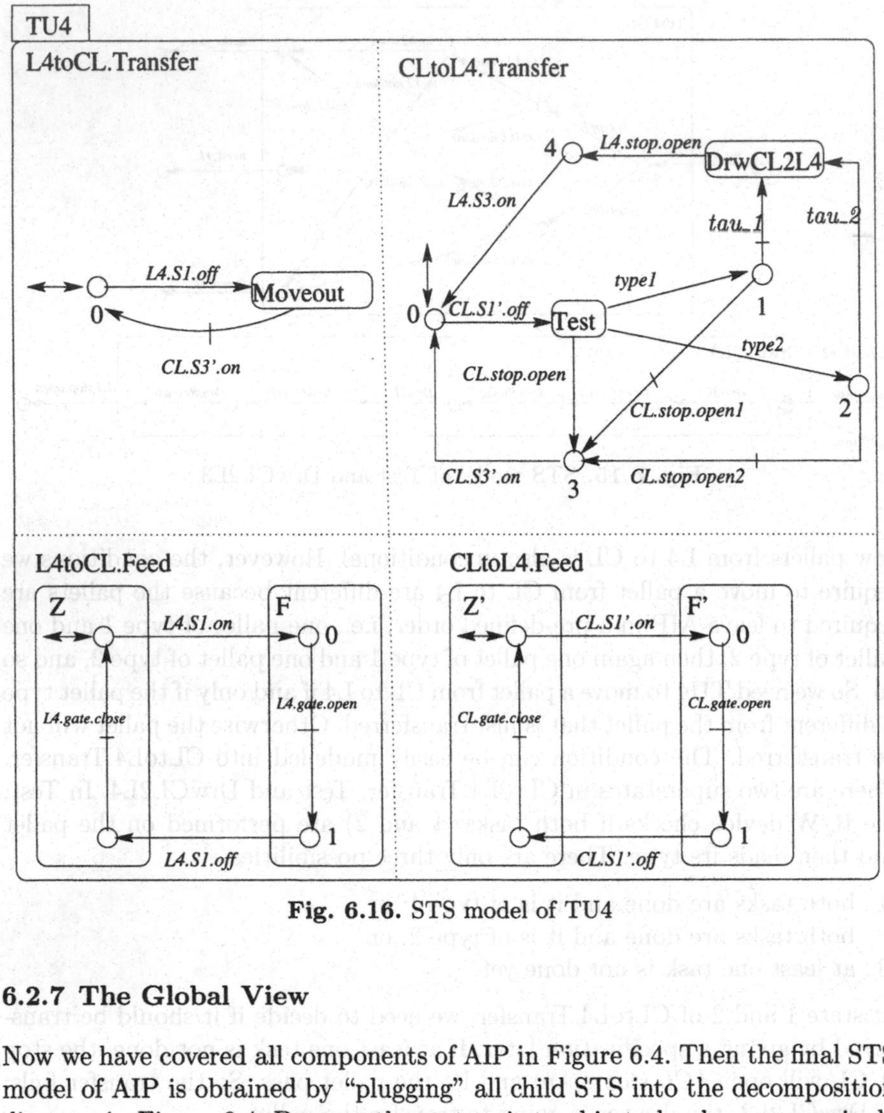

Fig. 6.16. STS model of TU4

6.2.7 The Global View

Now we have covered all components of AIP in Figure 6.4. Then the final STS model of AIP is obtained by "plugging" all child STS into the decomposition diagram in Figure 6.4. Because the system is too big to be drawn on a single page, we can draw the overall picture in Figure 6.20 and then refer the detailed dynamics of each subsystem to its STS model given in previous subsection. For example, in the overall picture, the detailed dynamics of AS1 is referred to Figure 6.5 on page 149.

The state tree of AIP is illustrated in Figure 6.21. The depth of this state tree is 5.

As far as synthesis is concerned, the most difficult part of this model is that the AIP's state space size (the number of basic state trees in AIP) is of order 10^{24}, much beyond the capacity of a synthesis tool based on extensional data representation.

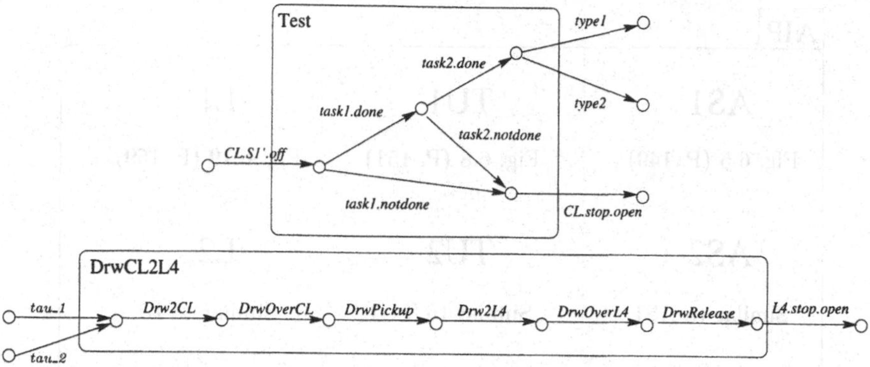

Fig. 6.17. STS model of Test and DrwCL2L4

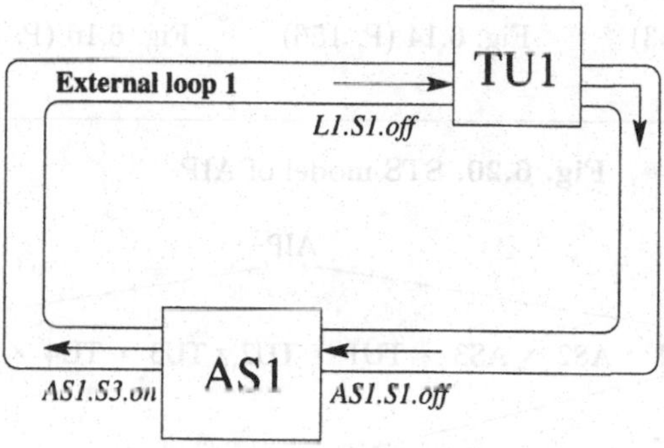

Fig. 6.18. Input/output of L1

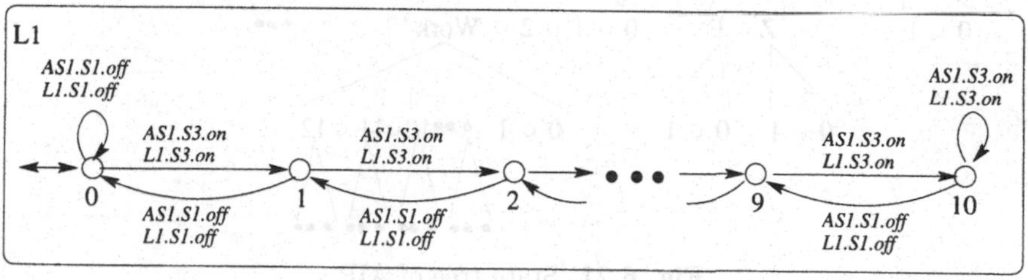

Fig. 6.19. STS model for the capacity of L1

6.3 Specifications

6.3.1 AS1, AS2

There are two requirements for AS1.

1. No buffer overflow. That is, only feed one pallet at a time to AS1. In other words, *no* event $AS1.S1.off$ is allowed to happen on the following states of AS1.Process: 1, 2 and AS1.Work. From the previous chapter, we can write a predicate for this *illegal* situation:

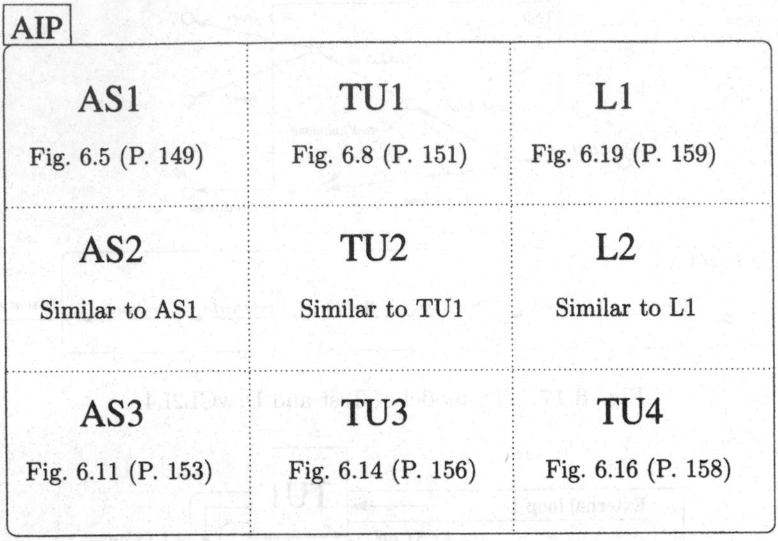

Fig. 6.20. STS model of AIP

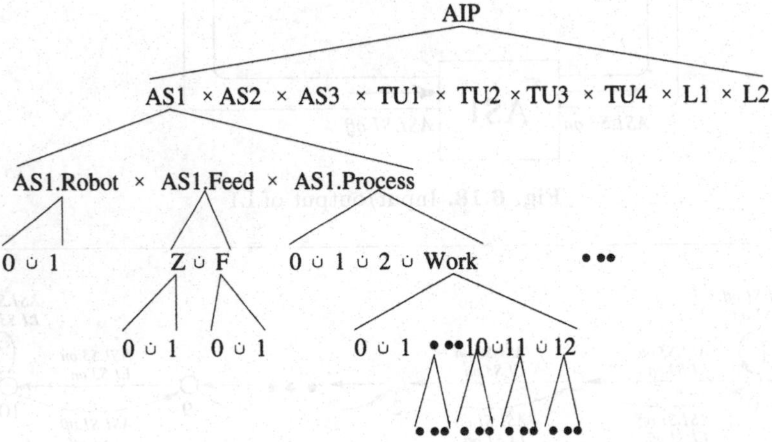

Fig. 6.21. State tree of AIP

$$v_{\mathrm{AS1.Feed}} = F \wedge v_F = 1 \wedge v_{\mathrm{AS1.Process}} \in \{1, 2, \mathrm{AS1.Work}\},$$

based on the STS model in Figure 6.5.

2. Choose *passing* (i.e., *AS*1.*stop.open*) when the robot is down and *processing* (i.e., *AS*1.*stop.close*) when it is on. Because both events are controllable, this requirement can be enforced by simply updating the holon AS1.Robot in Figure 6.5 by the following one. Notice that we relabel the event to *AS*1.*stop.open′* at the superstate AS1.Work of AS1.Process in Figure 6.5. The relabelling is necessary to make sure the pallet stop can

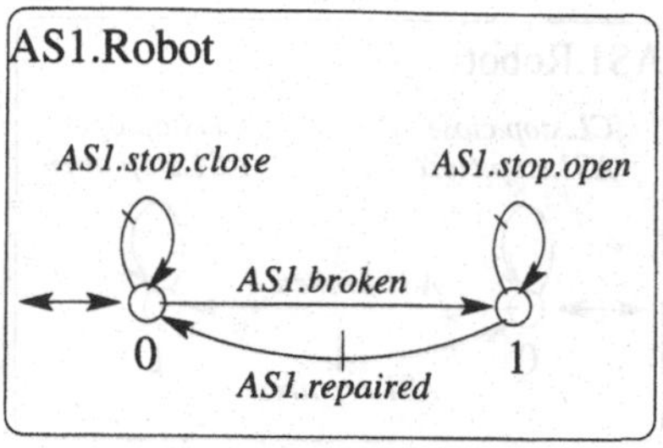

Fig. 6.22. Updated AS1.Robot

be opened when the robot is working properly (while $AS1.stop.open$ is disabled by the above specification).

The specification of AS2 is exactly the same as that of AS1 except state/event relabelling.

6.3.2 TU1, TU2

There are three requirements for TU1.

1. Mutual exclusion. It is not allowed to have one pallet fed from L1 and another one fed from CL *at the same time*. That is, the dangerous situation is when L1toCL.Transfer is in the superstate Moveout while CLtoL1.Transfer is at any states except the initial state 0. The predicate for this illegal situation is given by

$$v_{\mathrm{L1toCL.Transfer}} = \text{Moveout} \wedge v_{\mathrm{CLtoL1.Transfer}} \in \{1, 2, 3, \text{Test\&Move}\}.$$

2. No buffer overflow. That is, one pallet at a time should be fed from L1 or from CL. In other words, no event $L1.S1.off$ is allowed to happen on the superstate Moveout of L1toCL.Transfer and no event $CL.S1'.off$ is allowed to happen on the following states of CLtoL1.Transfer: 1, 2, 3 and Test&Move. We can write a predicate for this *illegal* situation:

$$v_{\mathrm{L1toCL.Feed}} = F \wedge v_F = 1 \wedge v_{\mathrm{L1toCL.Transfer}} = \text{Moveout} \vee$$
$$v_{\mathrm{CLtoL1.Feed}} = F' \wedge v_{F'} = 1 \wedge v_{\mathrm{CLtoL1.Transfer}} \in \{1, 2, 3, \text{Test\&Move}\}.$$

3. Transfer pallet from CL to L1 only if AS1 is on. This requirement can be enforced by simply updating the holon AS1.Robot in Figure 6.22 to the following one.

The specification of TU2 is exactly the same as that of TU1 except for state/event relabelling.

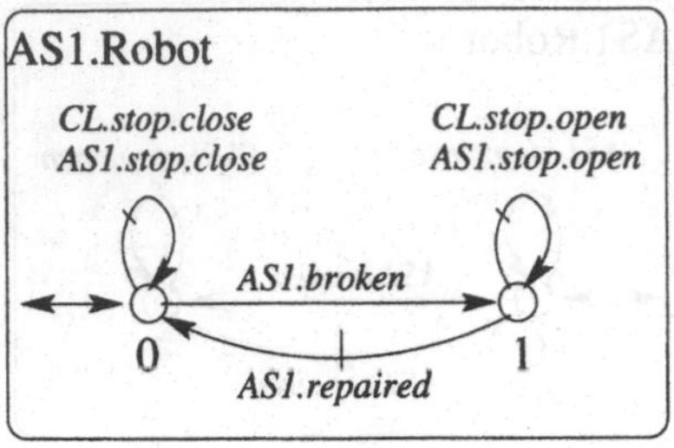

Fig. 6.23. Updated AS1.Robot

6.3.3 AS3

There are two requirements for AS3.

1. No buffer overflow. That is, one pallet at a time should be fed to AS3. In other words, *no* event *AS3.S1.off* is allowed to happen on any state of AS3.Process except the initial state 0. So it is easy to have $(\mathcal{S}', \textit{AS3.S1.off})$ as the specification, where

$$\mathcal{S}' := v_{\textbf{AS3.Process}} \neq 0.$$

Also, according to $\text{Elig}_{\text{AIP}}(\textit{AS3.S1.off})$, we can directly write a predicate for this *illegal* situation:

$$\mathcal{S} := v_{\textbf{AS3.Feed}} = F \wedge v_F = 1 \wedge v_{\textbf{AS3.Process}} \neq 0.$$

2. Choose *passing* or *processing* based on the status of AS1 and AS2. Basically, besides the case that AS3 must handle the faultily assembled pallets, there are four cases:
   a) If both AS1 and AS2 are working properly, we should let all unfinished pallets on L3 to pass (allowing *AS3.stop.open12* in AS3.Process) AS3 and send them back to be processed by AS1 or AS2.
   b) If only AS1 is broken down, we should process the pallets that need to do task1 (allowing *AS3.read, AS3.tau1*) but send the pallets that need to do task2 back to AS2 (allowing *AS3.stop.open2*).
   c) If only AS2 is broken down, we should process the pallets that need to do task2 (allowing *AS3.read, AS3.tau2*) but send the pallets that need to do task1 back to AS1 (allowing *AS3.stop.open1*).
   d) If both AS1 and AS2 are down, we have to let AS3 perform both tasks on the pallets in L3 (allowing *AS3.read, AS3.tau1, AS3.tau2*).

   Because all events mentioned here are controllable, the requirements can be applied by simply adding a memory (specification holon) AS3.Robot as a child holon of AIP, shown in Figure 6.24.

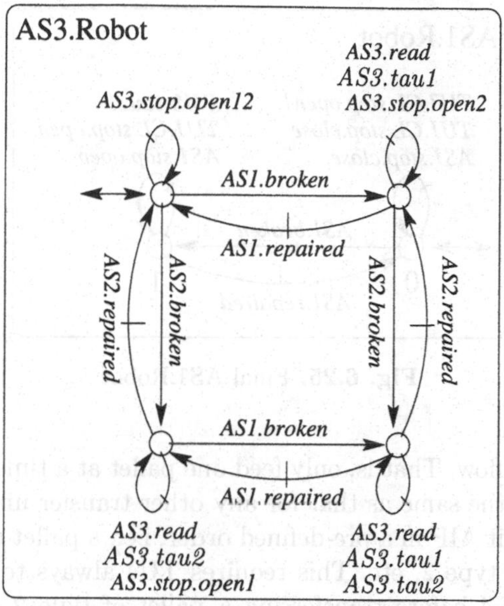

Fig. 6.24. Holon AS3.Robot

6.3.4 TU3

There are three requirements for TU3.

1. Mutual exclusion. It is not allowed to have one pallet from L3 and another one from CL to enter TU3 *at the same time.* This specification is the same as that for TU1 and TU2.
2. No buffer overflow. That is, only one pallet at a time should be fed from L3 or from CL. This is also the same as that for TU1 and TU2.
3. Transfer unfinished pallets from CL to L3 if either AS1 or AS2 is down. We just need to look at the case when AS1 is down, and handle the other case similarly. This requirement can be enforced by simply updating the holon AS1.Robot in Figure 6.23 again to the one in Figure 6.25. The only changes are disabling the event *TU3.CL.stop.open1* at state 1 of AS1.Robot (when AS1 is broken down) to transfer the pallet into L3, and disabling the event *TU3.tau_1* at state 0 (when AS1 is working properly) to open the stop gate and let the pallet pass.

6.3.5 TU4

There are three requirements for TU4.

1. Mutual exclusion. It is not allowed to have one pallet from L4 and another one from CL *at the same time.* This specification is same as that for the other transfer units.

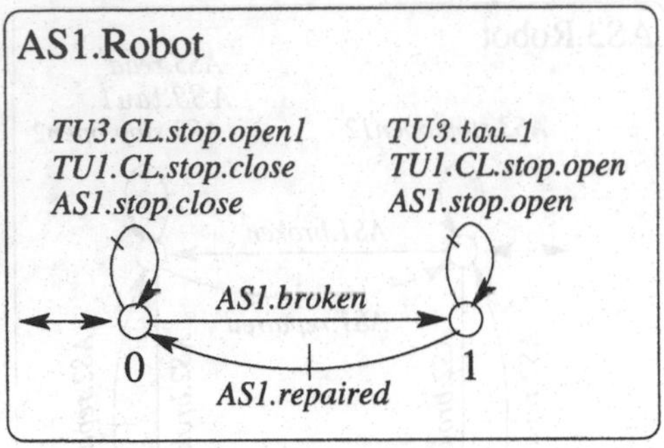

Fig. 6.25. Final AS1.Robot

2. No buffer overflow. That is, only feed one pallet at a time from L4 or from CL. It is also the same as that for any other transfer units.
3. The pallets exit AIP in a pre-defined order, i.e., a pallet of type 1 followed by a pallet of type 2, etc. This requires TU4 always to transfer a pallet of type 1 to L4 before transferring a pallet of type 2. We describe the specification by adding a new child holon CLtoL4.Exit for TU4, shown in Figure 6.26. At state 0, all pallets of type 2 are allowed to pass. The arrival of a pallet of type 1 leads to state 1, where we keep the stop gate closed (disabling *CL.stop.open1*) to transfer the pallet to L4 (event *tau_1*). Then the arrival of a pallet of type 2 will lead to state 3, where it is transferred to L4.

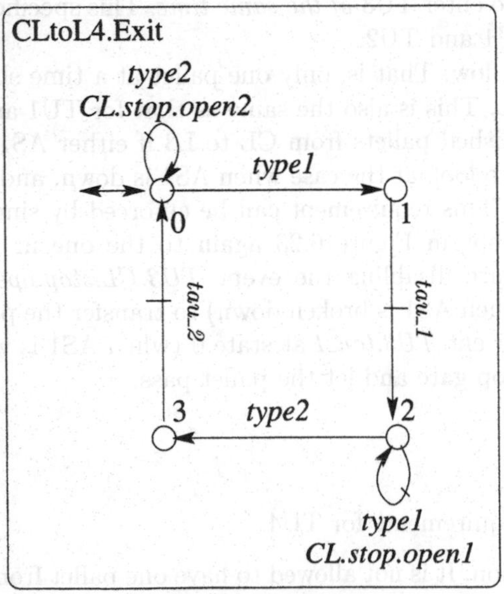

Fig. 6.26. Sequencing of pallets on exit

Notice that all events being disabled in CLtoL4.Exit are controllable. So it is a valid memory (specification holon).

6.3.6 L1, L2

At most 10 pallets are allowed to enter each external loop. As illustrated in Figure 6.19, the events *AS1.S3.on* and *L1.S3.on* are not allowed to occur at state 10 of the holon L1. As both events are uncontrollable, we need to write the logic specification as $(\mathcal{S}, AS1.S3.on)$ and $(\mathcal{S}, L1.S3.on)$, where

$$\mathcal{S} := v_{\text{L1}} = 10.$$

Similarly, we can write the logic specification for the capacity of L2.

6.4 Nonblocking Supervisory Control of AIP

The AIP example looks very complex, with the state space of order 10^{24} and 19 logic formulas as its specification. However, our BDD-based program can automatically compute the controller in less than 20 seconds, using a fixed allocation of memory (required by the BDD package we use) on a personal computer with 1G Hz Athlon CPU and 256MB RAM. The resulting controlled behavior, i.e., the BDD for the set of legal basic state trees, has 717 BDD nodes.

The AIP example has 28 OR superstates that are AND-adjacent to the root state in the top level. The good news is that we can divide them into 4 *clusters*, with no events shared between any two of them. Then the improved algorithms in Chapter 4 can be applied to achieve efficiency. The bad news is that one cluster has 21 AND components, still very big to handle. We list the clusters in the Table 6.1. Notice that even though there are no events shared among these clusters, we cannot divide the given control problem into

Table 6.1. Clusters of the AND root state in AIP

Cluster #	Elements in the Cluster
1	AS1.Robot AS1.Feed AS1.Process AS2.Robot AS2.Feed AS2.Process AS3.Robot AS3.Feed AS3.Process TU1.L1toCL.Feed TU1.L1toCL.Transfer TU1.CLtoL1.Feed TU1.CLtoL1.Transfer TU2.L2toCL.Feed TU2.L2toCL.Transfer TU2.CLtoL2.Feed TU2.CLtoL2.Transfer TU3.CLtoL3.Feed TU3.CLtoL3.Transfer L1 L2
2	TU3.L3toCL.Feed TU3.L3toCL.Transfer
3	TU4.L4toCL.Feed TU4.L4toCL.Transfer
4	TU4.CLtoL4.Feed TU4.CLtoL4.Transfer TU4.CLtoL4.Exit

4 independent problems among the 4 clusters, because, for example, there are specifications on both the states in TU3.CLtoL3.Transfer of cluster 1 and the states in TU3.L3toCL.Transfer of cluster 2 (the mutual exclusion specification).

Table 6.2 lists the BDD size of the control functions ($f_\sigma = \Gamma(N_{good}, \sigma)$, see chapter 4 for the section: control implementation) for some critical controllable events. All other controllable events are enabled everywhere, which means they are allowed to happen as long as they are eligible at the given basic state tree.

Each control function is simple, with the largest one (wrt. the number of BDD nodes) having only 17 nodes. Figure 6.27 is the control function of the event *AS1.gate.open*. It has only 2 BDD nodes. Because there are 4 states in

Table 6.2. Control functions of AIP

Controllable event σ	BDD size of f_σ
AS1.gate.open	2
AS1.stop.close	17
AS1.stop.open	17
AS2.gate.open	2
AS2.stop.close	17
AS2.stop.open	17
AS3.gate.open	5
TU1.L1.gate.open	6
TU1.CL.gate.open	6
TU1.CL.stop.close	6
TU2.L1.gate.open	6
TU2.CL.gate.open	6
TU3.CL.stop.close	6
TU3.L1.gate.open	7
TU3.CL.gate.open	7
TU4.L1.gate.open	6
TU4.CL.gate.open	6

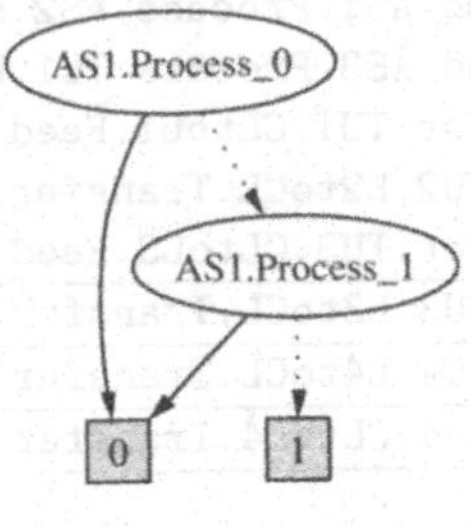

Fig. 6.27. Control function of *AS1.gate.open*

AS1.Process, we need 2 binary variables to encode them. We choose to encode the states in AS1.Process with the Least Significant Bit AS1.Process_0 first in the ordering (top of the BDD). Each node has two edges. The dotted line is with value 0 and the solid line with value 1. We can translate the graph into the following rule:

enable *AS1.gate.open* if and only if the holon AS1.Process is at state 0.

See how simple it is! Investigating the AIP specifications, the reader will find that this control rule is necessary to avoid taking two pallets into AS1 (buffer overflow).

Figure 6.28 is the control function of the event *TU1.CL.gate.open*. We can also translate it into two simple rules. That is, we enable *TU1.CL.gate.open* if and only if one of the following conditions holds,

1. TU1.L1toCL.Feed at 0, TU1.L1toCL.Transfer at 0, and TU1.CLtoL1. Transfer at 0;
2. TU1.L1toCL.Feed at 1, TU1.L1toCL.Feed.1 at 0, TU1.L1toCL.Transfer at 0, and TU1.CLtoL1.Transfer at 0.

One of the largest control functions listed in Table 6.2 is given in Figure 6.29. It is slightly complicated, but still not difficult to follow what it is doing.

Notice that an electronic computer is better adapted to the BDD representation of the control functions, as the computational complexity of supervision is linear in the number of BDD nodes. But the human user may well prefer a set of control rules. Our program can generate both. A better alternative might be to apply IDD (Integer Decision Diagram) to our approach, because

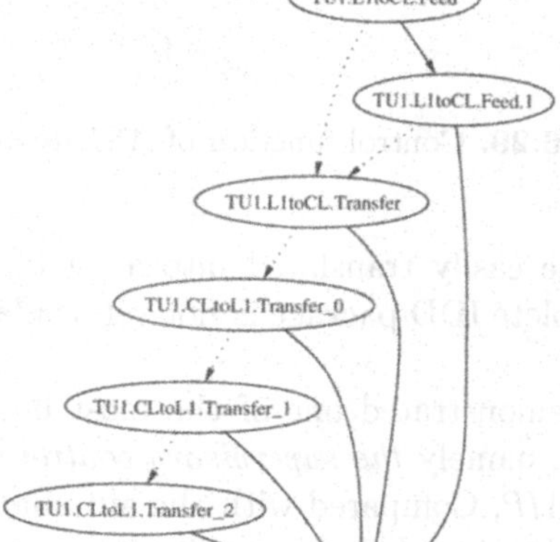

Fig. 6.28. Control function of *TU1.CL.gate.open*

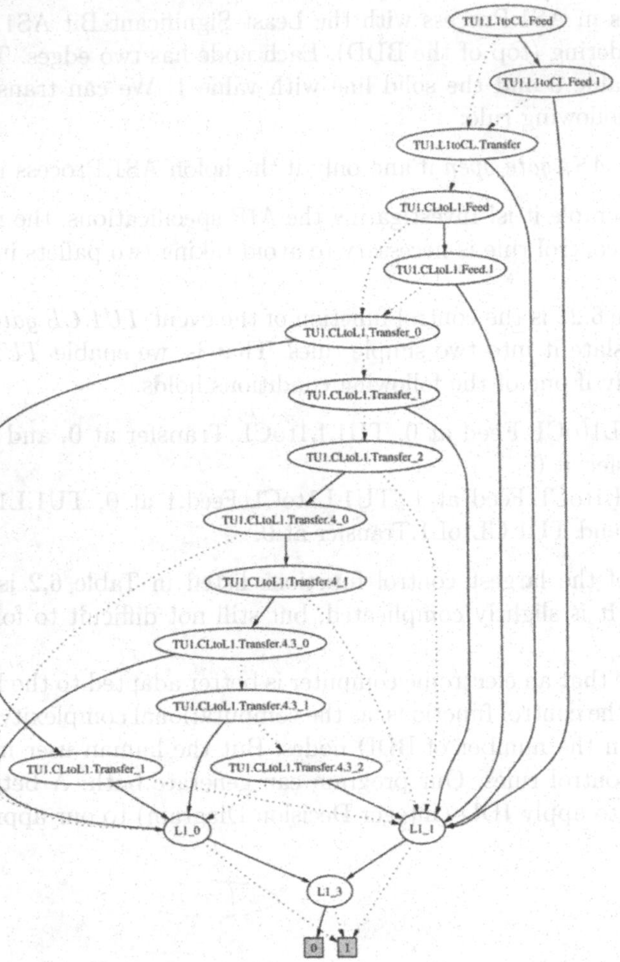

Fig. 6.29. Control function of *AS1.stop.close*

an IDD graph is quite easily translated into a set of understandable rules. Unfortunately, a complete IDD package is not yet available at the time of this research.

So far, we have demonstrated one of the most beautiful (we think) features of our approach, namely *the supervisory control logic can be simple for a system as large as AIP*. Compared with the automaton controller that has complexity linear in the size of the system's state space, our logic-based, or rule-based, control functions make RW supervisory control much more attractive for industrial applications.

6.5 Summary

The AIP example demonstrates the modelling and synthesis power of our STS framework. It is easy to understand the dynamics of AIP from our STS model as well as the specifications written by predicates. The synthesis result is better than that obtainable by any other currently available synthesis tools in terms of time and space. Finally, the correctness of the resulting controller is guaranteed.

7

Conclusions and Future Research

7.1 Summary

In this book, we introduced a state-based version of Wang's State Tree Structure (STS) to manage complex system behavior. In order to perform control design efficiently, we employed a symbolic representation of STS and developed a recursive symbolic algorithm that succeeds with the AIP example having state space of order 10^{24}. The state space explosion problem is effectively controlled. Furthermore the resulting controllers are tractable and highly comprehensible.

First of all, STS is a modelling tool. It is a natural extension of Finite State Machines (FSM) and Synchronous Product Systems (SPS). Here we summarize some important properties of STS.

1. Its multilevel hierarchical structure can put local information (i.e., transitions labelled with unique events) in lower levels and coupling information (with shared events) bounded under one AND superstate.
2. In general, the initial state tree has more than one basic sub-state-tree. This relaxation of the conventional initialization requirement provides greater modelling flexibility.
3. Unlike in an automaton where the transition graph is laid out explicitly, the *intensional* definition of the transition function Δ makes it possible to store a complex system's transition graph in the computer *implicitly*.

Finite state machines (FSM) and Synchronous Product Systems (SPS) are special STS. So trivially, any control problems described by FSM or SPS can be rewritten in our STS setting. The usefulness of STS, as we see it, lies in its modularity (including structured state space and local coupling). So we want to retain the modularity of STS at all times.

When an STS model is developed initially, the control designer will probably have certain types of specifications in mind, and that may well govern the way components and subcomponents are placed on the state tree. In other words, it is not as though the STS model were created first, independently of any notion of specification, and then the specifications were brought in afterwards. So in fact, once the STS model is in place, the apparent formal restriction on what specifications are possible should not, in practice, be so restrictive.

As long as a system can be modelled as a STS, we have developed a complete symbolic approach to design its controller. We highlight a few important properties of our approach.

1. The symbolic representation of STS is decentralized, i.e., no centralized transition relation is needed. Accordingly the computation of Γ is efficient.
2. The depth-first recursive algorithm takes advantage of our encoding scheme and the hierarchical structure of STS. It effectively "controls" the BDD size of intermediate predicates.
3. The simplified control functions make our control implementation transparent and tractable.

Finally, our synthesis approach can be directly applied to nondeterministic STS models. The only question is how to encode nondeterministic transitions, which we will explain in the section on future work.

7.2 Discussion on Computational Complexity

In [GW00], it is proven that, for the synchronous product systems, the nonblocking supervisory control problem is NP-hard, i.e, *in the worst case* (and assuming $P \neq NP$), any synthesis algorithm's computational complexity will inevitably be measured by the model's state size, which is exponential in the number of system components. This statement also applies to our STS framework, because synchronous product systems are just special STS.

The original TCT program computes the global plant automaton every time before the synthesis, which unfortunately turns every run of the synthesis into the worst case. However, we believe that *the worst case may actually not occur in many practical control problems if we take advantage of the structural information embedded in our STS model,* e.g., the AIP example.

Our computer program works extremely well for those practical problems to which our STS setting is adapted, especially those with comparatively few shared events. It can also work well for some systems with all of their events shared among components. Basically, two essential factors contribute to its computational efficiency. One is the symbolic computation based on BDD, the other is STS structural information.

7.2.1 Contributions of BDD-based Symbolic Computation

BDD are efficient representations of predicates, or subsets of the model's state space. Already in chapter 1 it was observed that, after introducing BDD to represent predicates, the synthesis' computational complexity is polynomial in the number of BDD nodes in use, instead of the model's state size.

After giving our algorithms in chapter 4, we can summarize the contributions of the BDD approach in the following respects.

1. After encoding, the synthesis' computational complexity is measured by the number of BDD nodes in use. With a suitable ordering of variables, this number is much smaller than the model's state size.
2. The fixpoint computation benefits from BDD because the equivalence checking of two BDD costs constant time in our BDD package.
3. The tautology $(\bigvee_{y \in \mathcal{E}(x_o)} (v_{x_o} = y)) \equiv 1$ is *automatically* enforced in our BDD package to simplify those intermediate predicates during the computation.[1]

However, without structural information, using BDD alone cannot handle large systems comfortably. Without structure, a flat automaton plant is a special STS (with one holon). Call it the *trivial* STS, and an STS that has more than one holon *nontrivial*. We can encode the trivial STS state space and transition graph using the same Θ function as in chapter 4. However, in this case, various technical problems will limit BDD's capability of handling complex systems.

1. For a complex system with a huge number of states, it is difficult to encode its *transition graph*. This encoding is at least linear in the model's state size (we have to encode the transitions one by one). Notice that the encoding is a part of the synthesis. So the synthesis' computational complexity will be measured by the model's state size!
2. There is only one state variable v, with its range over the entire state space. To apply BDD, we need to encode v by a large number of boolean variables. For a flat complex model, it is nearly impossible to find a heuristic approach which yields an improved ordering of these boolean variables.
3. Because of the centralized encoding of the state variable v, the transition relation given in chapter 4, labelled by a fixed event σ, may need many BDD nodes, simply because each transition from a state p to another state q requires *all* boolean variables to encode both p and q.[2]
4. The tautology $(\bigvee_{y \in \mathcal{E}(x_o)} (v_{x_o} = y)) \equiv 1$ cannot help in speeding up the computation, again because of the single state variable v.
5. There is no clear guide to choose the best computing direction.[3]

The above are the main reasons that the flat algorithm in chapter 4 did not succeed in the AIP example.

---

[1] This is from the fact that all BDD are *reduced*. Refer to [Bry86].

[2] For a nontrivial STS, such transition relation usually depends on *a subset* of the set of all boolean variables, which usually means a smaller number of BDD nodes. Refer to chapter 4 for detail.

[3] The nontrivial STS contributes to our depth-first algorithm.

7.2.2 Contributions of STS

With nontrivial STS (i.e., ruling out flat automata, which are trivial STS), all of the above problems can be readily overcome.

1. With nontrivial STS, it is easy to encode the transition graph, block by block, from the set of holons. The encoding for each given event σ is linear in the number of transitions labelled with σ in *the STS model*, which is much smaller than the number of transitions in *the equivalent flat automaton model*.
2. With nontrivial STS, we find a good ordering of BDD variables for our algorithms. For example, if holon A is the parent of holon B, then the BDD variables of holon A should precede those of holon B.[4] After incorporating this ordering in our AIP example, the new computer program takes only about one hundredth of the time of the original program.
3. With nontrivial STS, the transition relation of the given event σ usually needs a rather small number of BDD nodes, because σ usually occurs in a small set of holons. Then *only the state variables of those holons and their common ancestor holons* will appear in the transition relation, which of course requires a smaller number of BDD nodes, compared with the encoding of unstructured systems which requires *all* boolean variables. The greater simplicity of the transition relations can further contribute to the fast computation of Γ.
4. With nontrivial STS, the tautology $(\bigvee_{y \in \mathcal{E}(x_o)} (v_{x_o} = y)) \equiv 1$ can now be applied to simplify intermediate BDD during the synthesis.
5. With nontrivial STS, we finally give our depth-first algorithms that *guide* the entire computation. We believe that choosing the best computation direction is the key of the synthesis. This is also the main reason that our structured algorithms can handle the AIP example.

In summary, symbolic computation is an essential part of our computation. It provides the basic means for us to explore the structures in our STS model, especially the AND structure. Besides BDD, we could also choose other decision diagrams such as IDD (Integer Decision Diagrams) as the symbolic tool. However, it is the STS that provide the guidance to better computation. In our program, it is the STS that suggest a better ordering of BDD variables, a better encoded transition relations, and a direction for fast computation.

[4] The reason is from the encoding function Θ, i.e., the state variable of holon B is *always* in conjunction with the variable of holon A. Then in a BDD graph, knowing the assignment of holon A first will make the decision-making faster, consequently a smaller number of BDD nodes is required.

7.3 Future Work

7.3.1 Extended STS Model

A superstate is a *representative* of all of its descendants. Some *common* properties of its descendants can therefore be given easily in terms of the superstate.

One is illustrated in Figure 7.1. The selfloop of superstates is not allowed in our STS model because it is hard to define its meaning. However, if we define the selfloop of event τ in Figure 7.1 to mean that every descendant of superstate 9 has a selfloop of τ, we can encode this group of transitions by

$$v'_x = 9 \wedge v_x = 9.$$

One then finds that the symbolic computation of Γ introduced in this book can still be applied to this transition formula without any change. Therefore, our synthesis program can still work on this extended STS model.

Another example is given in Figure 7.2. For each given event τ, we only allow a single transition of τ to leave a superstate in our STS model. However, if we define the transition in Figure 7.2 to mean that every descendant of superstate 9 has a transition to the simple state 10, we can encode this group of transitions by

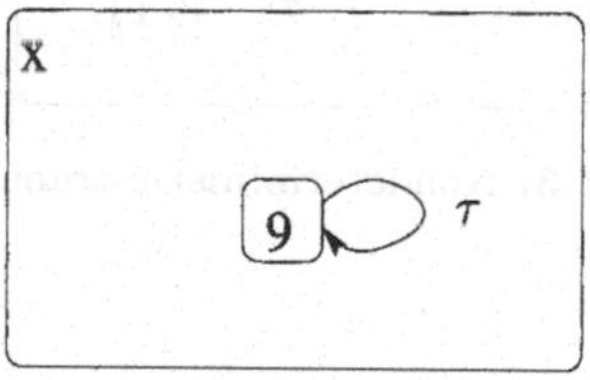

Fig. 7.1. Selfloop of superstates

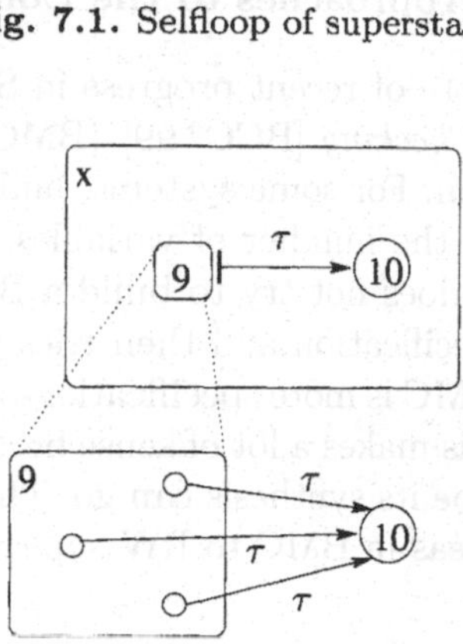

Fig. 7.2. Group transitions from a superstate to a simple state

$$v'_x = 9 \wedge v_x = 10.$$

Again, our symbolic computation of Γ is still valid.

So we can extend our STS model to include the above two types of transitions, without any need to change our synthesis approach.

For a nondeterministic DES, with transitions illustrated in Figure 7.3, we can encode the nondeterministic transitions of τ by

$$v'_x = 9 \wedge v_x \in \{10, 11\}.$$

Our symbolic computation of Γ remains valid in this case, too.

A more fundamental extension of our STS model could be to allow the boundary and inner transition graphs of each superstate to be represented by other means as Petri Nets or logic formulas, besides holons.

Another fundamental extension of our STS model is to allow events shared at different levels. This way it will be much easier to model a complex system and write a memory (specification holon). However, we pay the price in computation because it may be much more intricate.

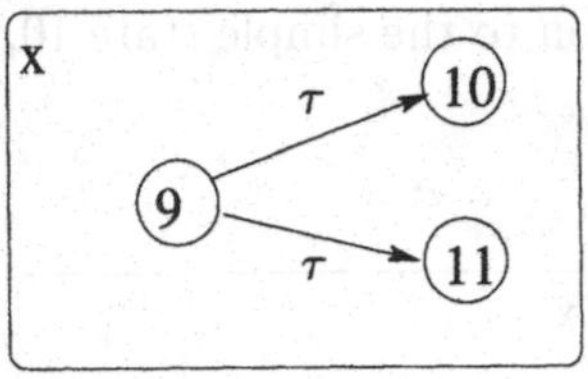

Fig. 7.3. Nondeterministic transitions

7.3.2 Other Symbolic Approaches to the Control of DES

It is interesting to take note of recent progress in Symbolic Model Checking (SMC). *Bounded Model Checking* [BCC+99] (BMC) without using BDD is attracting a lot of attention. For some systems, building the transition graph by BDD is exponential in the number of variables, no matter what order of variables is chosen. BMC does not try to build a BDD, but instead builds a logic formula from the specification and then tries to *prove* it by using some satisfiability provers. So BMC is more specification-oriented and therefore very efficient in some cases. This makes a lot of sense because the easier the control problem, the faster we hope its synthesis can go. Thus, it will be appealing to see if we can adapt the ideas in BMC to RW supervisory control theory.

7.3.3 Suboptimal Control of DES based on Symbolic Computation

Sometimes our design goal is just to find a feasible controller, ensuring controllability and nonblocking. This could open up more options. For example, we may just need to compute $B \preceq CR(G, P)$ having a smaller number of BDD nodes, instead of computing $CR(G, P)$ itself. Furthermore, it may not be necessary to compute a fixpoint, which is computationally expensive.

7.3.4 Application of Symbolic Approach to Other DES Areas

In standard theory [Won04], to compute a supervisor with partial observations needs exponential effort (rather than polynomial) in the model's state size, owing to the necessity of converting from a nondeterministic automaton to an equivalent deterministic automaton [Won04]. But our approach can be applied to nondeterministic STS model without transforming it to a deterministic model. So it may help in computing supervisors with partial observations.

It will also be of interest to apply the symbolic approach to Timed DES [BW94], fair supervision [Goh03], and other potential extensions of SCT.

This book is an attempt to combine RW supervisory control theory (SCT), statecharts, and symbolic computation together. While the results are highly encouraging, they are still at an early stage. Research is needed on our STS framework to explore further structural and dynamic properties, experiment with more symbolic computational ideas, and develop more powerful synthesis algorithms.

References

[And97] Henrik Reif Andersen. *An Introduction to Binary Decision Diagrams.* Lecture notes for 49285 Advanced Algorithms E97. Department of Information Technology, Technical University of Denmark `http://www.it.dtu.dk/~hra`, 1997.

[Arn88] D. Arnon. A bibliography of quantifier elimination for real closed fields. *Journal of Symbolic Comput.*, 5(1-2):267–274, 1988.

[BC94] B. Brandin and F. Charbonnier. The supervisory control of the automated manufacturing system of the AIP. In *Proc. Rensselaer's 1994 4th Intl. Conf. on Computer Integrated Manufacturing and Automation Technology*, pages 319–324, Troy, 1994.

[BCC98] Sergey Berezin, Sérgio Campos, and Edmund M. Clarke. Compositional reasoning in model checking. *Lecture Notes in Computer Science*, 1536:81–102, 1998.

[BCC+99] A. Biere, A. Cimatti, E.M. Clarke, M. Fujita, and Y. Zhu. Symbolic model checking using SAT procedures instead of BDDs. In *Design Automation Conference, (DAC'99)*, June, 1999.

[BCM+92] J.R. Burch, E.M. Clarke, K.L. McMillan, D.L. Dill, and L.J. Hwang. Symbolic model checking: 10^{20} states and beyond. *Information and Computation*, 98:142–170, June 1992.

[BH93] Y. Brave and M. Heymann. Control of discrete event systems modeled as hierarchical state machines. *IEEE Transactions on Automatic Control*, 38(12):1803–1819, December 1993.

[BHG+93] S. Balemi, G.J. Hoffmann, P. Gyugyi, H. Wong-Toi, and G.F. Franklin. Supervisory control of a rapid thermal multiprocessor. *IEEE Transactions on Automatic Control*, 38(7):1040–1059, July 1993.

[BL00] G. Barrett and S. Lafortune. Decentralized supervisory control with communicating controllers. *IEEE Transactions on Automatic Control*, 45(9):1620–1638, September 2000.

[BLA+99] Gerd Behrmann, Kim G. Larsen, Henrik Reif Andersen, Henrik Hulgaard, and Jørn Lind-Nielsen. Verification of hierarchical state/event systems using reusability and compositionality. *TACAS: Tools and Algorithms for the Construction and Analysis of Systems, Lecture Notes in Computer Science*, 1579:163–177, 1999.

[Bra] Artur Brauer. Simulation of an industrial production cell. `http://www.fzi.de/divisions/prost/projects/production_cell/`.

[Bry86] R.E. Bryant. Graph-based algorithms for boolean function manipulation. *IEEE Trans. Computers*, 35(8):677–691, August 1986.

[BW94] B.A. Brandin and W.M. Wonham. Supervisory control of timed discrete-event systems. *IEEE Trans. Autom. Control*, 39(2):329–342, February 1994.

[CCM95] S. Campos, E.M. Clarke, and M. Minea. Verifying the performance of the PCI local bus using symbolic techniques. In *Proc. of the IEEE International Conference on Computer Design*, pages 73–79, 1995.

[CGH+93] E.M. Clarke, O. Grumberg, H. Hiraishi, S. Jha, D.E. Long, K.L. McMillan, and L.A. Ness. Verification of the futurebus+ cache coherence protocol. In *Proc. of the Eleventh International Symposium on Computer Hardware Description Languages and their Applications*, pages 5–20, North-Holland, April 1993.

[EGKP97] Hartmut Ehrig, Robert Geisler, Marcus Klar, and Julia Padberg. Horizontal and vertical structuring techniques for statecharts. In *International Conference on Concurrency Theory*, pages 181–195, 1997.

[FBHL84] A. A. Fraenkel, Y. Bar-Hillel, and A. Levy. *Foundations of Set Theory*. Elsevier Science Publishers B. V., 1984.

[GHL94] Ling Gou, Tetsuo Hasegawa, and Peter Luh. Holonic planning and scheduling for a robotic assembly testbed. In *Proc. of the fourth international conference on computer integrated manufacturing and automation technology*, pages 142–149, Oct. 10-12, 1994.

[Goh98] P. Gohari. A linguistic framework for controlled hierarchical DES. Master's thesis, Department of Electrical and Computer Engineering, Univ. of Toronto, 1998.

[Goh03] P. Gohari. *Fair Supervisory Control of Discrete Event Systems*. PhD thesis, Dept. of Electrical & Computer Engineering, University of Toronto, 2003.

[GRX03] A. Ghaffari, N. Rezg, and X. Xie. Feedback control logic for forbidden-state problems of marked graphs: Application to a real manufacturing system. *IEEE Transactions on Automatic Control*, 48(1):18–29, January 2003.

[Gun97] Johan Gunnarsson. *Symbolic Methods and Tools for Discrete Event Dynamic Systems*. PhD thesis, Linköping Studies in Science and Technology, 1997.

[GW98] Peyman Gohari and W. M. Wonham. A linguistic framework for controlled hierarchical DES. In *4th International Workshop on Discrete Event Systems (WODES '98), IEE,*, pages 207–212, 1998.

[GW00] Peyman Gohari and W. M. Wonham. On the complexity of supervisory control design in the RW framework. *IEEE Transactions on Systems, Man and Cybernetics, Special Issue on DES*, 30(5):643–652, 2000.

[Har87] D. Harel. Statecharts: A visual formalism for complex systems. *Science of Computer Programming*, 8:231–274, June 1987.

[HC02] P. Hubbard and P. E. Caines. Dynamical consistency in hierarchical supervisory control. *IEEE Transactions on Automatic Control*, 47(1):37–52, January 2002.

[Ho89] Y. C. Ho. Special issue on dynamics of discrete event systems. *Proceedings of the IEEE*, 77(1), January 1989.

[HT84] Dov Harel and Robert Endre Tarjan. Fast algorithms for finding nearest common ancestors. *SIAM Journal on Computing*, 13(2):338–355, May 1984.

[HWT92] G. Hoffmann and H. Wong-Toi. Symbolic synthesis of supervisory controllers. In *Proc. of 1992 American Control Conference*, pages 2789–2793, Chicago, IL, USA, 1992.

[Koe89] Arthur Koestler. *The Ghost in the Machine.* Penguin Group Ltd, 1989.

[LBW01] R.J. Leduc, B.A. Brandin, and W.M. Wonham. Hierarchical interface-based supervisory control: Serial case. In *Proc. of the 40th Conf. Decision Contr.*, pages 4116–4121, Dec. 4-7, 2001.

[Led02] R.J. Leduc. *Hierarchical Interface Based Supervisory Control.* PhD thesis, University of Toronto, Toronto, 2002.

[Lin] Thomas Lindner. Task description. `http://www.fzi.de/divisions/prost/projects/production_cell/`.

[LLW01a] R.J. Leduc, M. Lawford, and W.M. Wonham. *Hierarchical Interface Based Supervisory Control: AIP Example for Parallel Case.* Technical Report No. 2, Software Quality Research Laboratory, Dept. of Computing and Software, McMaster University, Hamilton, ON, Canada. `http://www.control.toronto.edu/~leduc/AIPreport.html`, 2001.

[LLW01b] R.J. Leduc, M. Lawford, and W.M. Wonham. Hierarchical interface-based supervisory control: Parallel case. In *Proc. of the 39th Allerton Conf. on Comm., Contr., and Comp*, pages 386 – 395, October 3-5, 2001.

[LLW01c] R.J. Leduc, M. Lawford, and W.M. Wonham. Hierarchical interface-based supervisory control: AIP example. In *Proc. of the 39th Allerton Conf. on Comm., Contr., and Comp*, pages 396 – 405, October 3-5, 2001.

[LW88] Y. Li and W.M. Wonham. Controllability and observability in the state-feedback control of discrete-event systems. In *Proc. of 27th IEEE Conf. Decision and Control*, pages 203–208, Austin, TX, USA, December 1988.

[LW90] F. Lin and W.M. Wonham. Decentralized control and coordination of discrete-event systems with partial observation. *IEEE Trans. Autom. Control*, 35(12):1330–1337, December 1990.

[LW93] Y. Li and W.M. Wonham. Control of vector discrete-event systems I – the base model. *IEEE Trans. Autom. Control*, 38(8):1214–1227, August 1993.

[Ma04] Chuan Ma. *Nonblocking Supervisory Control of State Ttree Structures.* PhD thesis, Department of Electrical and Computer Engineering, University of Toronto, 2004.

[MG02] H. Marchand and B. Gaudin. Supervisory control problems of hierarchical finite state machines. In *Proc. of the 41st IEEE Conf. on Decision and Control, Las Vegas, Nevada USA*, pages 1199–1204, 2002.

[Min02] R.S. Minhas. *Complexity Reduction in Discrete Event Systems.* PhD thesis, Dept. of Electrical & Computer Engineering, University of Toronto, 2002.

[MW03] R.S. Minhas and W.M. Wonham. Online supervision of discrete event systems. In *Proc. 2003 American Control Conference*, pages 1685–1690, June 2003.

[Ram83] P. J. Ramadge. *Control and Supervision of Discrete Event Processes.* PhD thesis, Department of Electrical Engineering, University of Toronto, 1983.

[RW82] P. J. Ramadge and W. M. Wonham. Supervision of discrete event processes. In *Proc. of 21st Conf. on Decision and Control*, pages 1228–1229, 1982.

[RW87a] P. J. Ramadge and W. M. Wonham. Supervisory control of a class of discrete event processes. *SIAM J. Contr. Optim.*, 25(1):206–230, 1987.

[RW87b] P.J. Ramadge and W.M. Wonham. Modular feedback logic for discrete event systems. *SIAM J. Control Optim.*, 25(5):1202–1218, September 1987.

[RW92] K. Rudie and W.M. Wonham. Think globally, act locally: Decentralized supervisory control. *IEEE Trans. Autom. Control*, 37(11):1692–1708, November 1992.

[SC02] G. Shen and P. E. Caines. Hierarchically accelerated dynamic programming for finite-state machines. *IEEE Transactions on Automatic Control*, 47(2):271–283, Feb 2002.

[Wan95] B. Wang. Top-down design for RW supervisory control theory. Master's thesis, Department of Electrical and Computer Engineering, Univ. of Toronto, 1995.

[Won04] W. M. Wonham. *Supervisory Control of Discrete-Event Systems.* Department of Electrical and Computer Engineering, University of Toronto `http://www.control.utoronto.ca/DES`, 2004.

[WR87] W.M. Wonham and P.J. Ramadge. On the supremal controllable sublanguage of a given language. *SIAM J. Control Optim.*, 25(3):637–659, May 1987.

[WR88] W.M. Wonham and P.J. Ramadge. Modular supervisory control of discrete event systems. *Mathematics of Control, Signals and Systems*, 1(1):13–30, 1988.

[WW96] K. C. Wong and W. M. Wonham. Hierarchical control of discrete-event systems. *Discrete-Event Dynamic Systems: Theory and Applications*, 6(3):241–273, July 1996.

[ZW90] H. Zhong and W.M. Wonham. On the consistency of hierarchical supervision in discrete-event systems. *IEEE Trans. Autom. Control*, 35(10):1125–1134, October 1990.

[ZW01] Z.H. Zhang and W.M. Wonham. Stct: An efficient algorithm for supervisory control design. In *Symposium on Supervisory Control of Discrete Event Systems (SCODES2001)*, Paris, July, 2001.

Index

List of Symbols

This is not an exhaustive list of special symbols. Only important ones are put in this list. Here R, X are state sets with $R \subseteq X$; x, y, z are states; $\mathbf{S}T$ is a state tree; H is a holon; $\mathbf{G}$ is a state tree structure; σ is an event; P is a predicate.

Symbol	Description	Page
$\mathcal{T}$	Type function of a given state	13
$\mathcal{T}_R$	Restriction of $\mathcal{T}$ to a subset R	13
$\mathcal{E}$	Expansion function of a given state	13
$\mathcal{E}_R$	Restriction of $\mathcal{E}$ to a subset R	13
$\mathcal{E}^*$	Reflexive and transitive closure of $\mathcal{E}$	14
$\mathcal{E}^+$	Unfolding of $\mathcal{E}$	14
$\mathbf{S}T^y$	A child state tree on $\mathbf{S}T$, rooted by y	14
$\mid$	$x \mid y$ if x and y are parallel	19
$\oplus$	$x \oplus y$ if x and y are exclusive	19
$\mathcal{S}T(\mathbf{S}T)$	Set of all sub-state-trees of $\mathbf{S}T$	21
$\mathcal{B}(\mathbf{S}T)$	Set of all basic sub-state-trees of $\mathbf{S}T$	31
δ_I	Internal transition function	34
δ_{BI}	Incoming boundary transition function	34
δ_{BO}	Outgoing boundary transition function	34
H^y	A holon matched to the superstate y	36
Δ	Transition function of $\mathbf{G}$	40
$\mathbf{G}^y$	A child STS of $\mathbf{G}$, with root state y	41
$\mathrm{Elig}_{\mathbf{G}}(\sigma)$	Largest eligible state tree of σ	47
Γ	Backward transition function of $\mathbf{G}$	54
$\mathrm{Next}_{\mathbf{G}}(\sigma)$	Largest target state tree of σ	54
$R(\mathbf{G}, P)$	Reachability subpredicate of P	62
$M_\sigma(P)$	Weakest liberal precondition of P	63
f	State feedback control (SFBC) for $\mathbf{G}$	63
f_σ	Control function of σ	63
$\mathbf{G}^f$	Controlled STS under SFBC f	63
$\mathcal{CP}(P)$	Family of weakly controllable subpredicates of P	66
$[P]$	A predicate transformer to compute $\sup \mathcal{CP}(P)$	67
$CR(\mathbf{G}, P)$	Coreachability subpredicate of P	69
$\mathcal{C}^2\mathcal{P}(P)$	Family of weakly contr. & coreachable subpredicates of P	71
Ω_P	A predicate transformer to compute $\sup \mathcal{C}^2\mathcal{P}(P)$	71
$\equiv$	Logical equivalence	83
Θ	Encoding function of sub-state-trees	84
$(N_\sigma, \mathbf{v}_{\sigma,\mathbf{S}}, \mathbf{v}_{\sigma,\mathbf{T}})$	Symbolic representation of the transitions labelled by σ	94
$\exists$	Existential quantifier	97
$\forall$	Universal quantifier	97
C	A predicate for the optimal controlled behavior	117

Lecture Notes in Control and Information Sciences

Edited by M. Thoma and M. Morari

Further volumes of this series can be found on our homepage: springeronline.com

Vol. 288: Taware, A. and Tao, G.
Control of Sandwich Nonlinear Systems
393 p. 2003 [3-540-44115-8]

Vol. 287: Mahmoud, M.M.; Jiang, J.; Zhang, Y.
Active Fault Tolerant Control Systems
239 p. 2003 [3-540-00318-5]

Vol. 286: Rantzer, A. and Byrnes C.I. (Eds)
Directions in Mathematical Systems
Theory and Optimization
399 p. 2003 [3-540-00065-8]

Vol. 285: Wang, Q.-G.
Decoupling Control
373 p. 2003 [3-540-44128-X]

Vol. 284: Johansson, M.
Piecewise Linear Control Systems
216 p. 2003 [3-540-44124-7]

Vol. 283: Fielding, Ch. et al. (Eds)
Advanced Techniques for Clearance of
Flight Control Laws
480 p. 2003 [3-540-44054-2]

Vol. 282: Schröder, J.
Modelling, State Observation and
Diagnosis of Quantised Systems
368 p. 2003 [3-540-44075-5]

Vol. 281: Zinober A.; Owens D. (Eds)
Nonlinear and Adaptive Control
416 p. 2002 [3-540-43240-X]

Vol. 280: Pasik-Duncan, B. (Ed)
Stochastic Theory and Control
564 p. 2002 [3-540-43777-0]

Vol. 279: Engell, S.; Frehse, G.; Schnieder, E. (Eds)
Modelling, Analysis, and Design of Hybrid Systems
516 p. 2002 [3-540-43812-2]

Vol. 278: Chunling D. and Lihua X. (Eds)
H_∞ Control and Filtering of
Two-dimensional Systems
161 p. 2002 [3-540-43329-5]

Vol. 277: Sasane, A.
Hankel Norm Approximation
for Infinite-Dimensional Systems
150 p. 2002 [3-540-43327-9]

Vol. 276: Bubnicki, Z.
Uncertain Logics, Variables and Systems
142 p. 2002 [3-540-43235-3]

Vol. 275: Ishii, H.; Francis, B.A.
Limited Data Rate in Control Systems with Networks
171 p. 2002 [3-540-43237-X]

Vol. 274: Yu, X.; Xu, J.-X. (Eds)
Variable Structure Systems:
Towards the 21st Century
420 p. 2002 [3-540-42965-4]

Vol. 273: Colonius, F.; Grüne, L. (Eds)
Dynamics, Bifurcations, and Control
312 p. 2002 [3-540-42560-9]

Vol. 272: Yang, T.
Impulsive Control Theory
363 p. 2001 [3-540-42296-X]

Vol. 271: Rus, D.; Singh, S.
Experimental Robotics VII
585 p. 2001 [3-540-42104-1]

Vol. 270: Nicosia, S. et al.
RAMSETE
294 p. 2001 [3-540-42090-8]

Vol. 269: Niculescu, S.-I.
Delay Effects on Stability
400 p. 2001 [1-85233-291-316]

Vol. 268: Moheimani, S.O.R. (Ed)
Perspectives in Robust Control
390 p. 2001 [1-85233-452-5]

Vol. 267: Bacciotti, A.; Rosier, L.
Liapunov Functions and Stability in Control Theory
224 p. 2001 [1-85233-419-3]

Vol. 266: Stramigioli, S.
Modeling and IPC Control of Interactive Mechanical
Systems – A Coordinate-free Approach
296 p. 2001 [1-85233-395-2]

Vol. 265: Ichikawa, A.; Katayama, H.
Linear Time Varying Systems and Sampled-data Systems
376 p. 2001 [1-85233-439-8]

Vol. 264: Baños, A.; Lamnabhi-Lagarrigue, F.;
Montoya, F.J
Advances in the Control of Nonlinear Systems
344 p. 2001 [1-85233-378-2]

Vol. 263: Galkowski, K.
State-space Realization of Linear 2-D Systems with
Extensions to the General nD (n>2) Case
248 p. 2001 [1-85233-410-X]

Vol. 262: Dixon, W.; Dawson, D.M.; Zergeroglu, E.;
Behal, A.
Nonlinear Control of Wheeled Mobile Robots
216 p. 2001 [1-85233-414-2]

Vol. 261: Talebi, H.A.; Patel, R.V.; Khorasani, K.
Control of Flexible-link Manipulators
Using Neural Networks
168 p. 2001 [1-85233-409-6]

Vol. 260: Kugi, A.
Non-linear Control Based on Physical Models
192 p. 2001 [1-85233-329-4]

Vol. 259: Isidori, A.; Lamnabhi-Lagarrigue, F.;
Respondek, W. (Eds)
Nonlinear Control in the Year 2000 Volume 2
640 p. 2001 [1-85233-364-2]

Vol. 258: Isidori, A.; Lamnabhi-Lagarrigue, F.;
Respondek, W. (Eds)
Nonlinear Control in the Year 2000 Volume 1
616 p. 2001 [1-85233-363-4]

Vol. 257: Moallem, M.; Patel, R.V.; Khorasani, K.
Flexible-link Robot Manipulators
176 p. 2001 [1-85233-333-2]

Printing and Binding: Strauss GmbH, Mörlenbach